CHINA:
Railways and Agricultural Development, 1875-1935

by

Ernest P. Liang

THE UNIVERSITY OF CHICAGO
DEPARTMENT OF GEOGRAPHY
RESEARCH PAPER NO. 203

1982

Copyright 1982 by Ernest P. Liang
Published 1982 by the Department of Geography
The University of Chicago, Chicago, Illinois

Library of Congress Cataloging in Publication Data

Liang, Ernest P., 1953–
   China, railways and agricultural development, 1875–1935.
   Research paper/University of Chicago, Department of Geography; no. 203)
   Bibliography: p. 169.
   1. Railroads—China—History. 2. Agriculture—Economic aspects—China—History. I. Title. II. Series: Research paper (University of Chicago. Dept. of Geography); no. 203.
H31.C514 no. 203 [HE3288] 910s [338.1'4] 82-4749
ISBN 0-89065-109-4   AACR2

Research Papers are available from:
The University of Chicago
Department of Geography
5828 S. University Avenue
Chicago, Illinois 60637
Price: $8.00; $6.00 series subscription

To My Wife

TABLE OF CONTENTS

LIST OF FIGURES vii

LIST OF TABLES ix

ACKNOWLEDGEMENT xi

Chapter

I. INTRODUCTION 1

II. THE DEVELOPMENT OF RAIL TRANSPORT IN CHINA 9

  Railway History to 1935: An Overview
  Rail Transport Development: Efficiency Aspects

III. RAILWAYS AND THE CHANGING PATTERNS OF DOMESTIC
  AGRICULTURAL TRADE 27

  The Growth of Interregional Trade
  The Spatial Configuration of Domestic Trade
    Expansion
  Railways and the Growth of Trade at Treaty Ports
  Summary

IV. THE IMPACT OF RAILWAYS ON AGRICULTURAL PRODUCTION . . . 61

  Introduction
  Market Accessibility and Agricultural Productivity:
    Hypotheses
  Empirical Testing of Hypothesis I
    (A) Farm Productivity and Market Accessibility
    (B) Allocative Efficiency and Accessibility
  Railways and the Growth of Agricultural Output
  The Effect of Railways on Land Values
  Summary

V. RAILWAY DEVELOPMENT AND AGRICULTURAL COMMERCIALIZATION . 97

  Introduction
  Regional Specialization and the Expansion of
    Cash Cropping: Evidence on Railway Effects
  Marketable Surplus and the Effect of Railways
  Summary

VI. TRANSPORT DEVELOPMENT AND RURAL ECONOMIC GROWTH:
  A GENERAL ASSESSMENT 125

  Introduction
  Trade Development and Rural Growth: Theory and
    Evidence
  Constraints on Railway Benefits and Their Impact
    on Chinese Agricultural Growth

VII. CONCLUSIONS 141

Appendix

A. ESTIMATION OF EXPECTED AVERAGE FARM OUTPUT 147

B. ESTIMATION OF THE ACCESSIBILITY INDEX 149

C. ESTIMATION OF THE VALUE OF MARGINAL PRODUCT
   PER MAN-WORKDAY AND DAILY LABOR COST 151

D. THE ENDOGENEITY AND EXOGENEITY OF RAILWAYS AS A FACTOR
   IN AGRICULTURAL CHANGES 155

E. ESTIMATION OF THE AVERAGE DECLINE IN RURAL-URBAN
   TRANSFER COSTS AND ITS CONTRIBUTION TO INCREASES
   IN FARM PRICES 165

BIBLIOGRAPHY 169

LIST OF FIGURES

1. Republican China: Political Regions of
   China Proper (c. 1930)   5

2. Railways in China, 1914 13

3. Railways in China, 1935 15

4. Long-distance Agricultural Trade in China, c. 1870 . . 34

5. Long-distance Agricultural Trade in China, c. 1935 . . 37-41

6. Railway and the Growth of Maritime Export Trade
   at Leading Treaty Ports, 1900-1930 50-52

7. Location of Sample <u>Hsien</u> (Counties) 67

8. Agricultural Regions in China 68

9. Location of Railway and Non-railway <u>Hsien</u> 90

10. Location of Railway and Non-railway <u>Hsien</u> in Honan with
    more than 1,000 <u>mou</u> in sesame (c. 1932) 105

11. Location of Railway and Non-railway <u>Hsien</u> in Honan with
    more than 1,000 <u>mou</u> in cotton (c. 1932) 107

12. Location of Railway and Non-railway <u>Hsien</u> in Shantung
    with more than 100 <u>mou</u> in tobacco (c. 1933) 110

13. Location of Railway and Non-railway <u>Hsien</u> in Hopei with
    more than 5,000 mou in cotton, peanuts, and sesame
    (c. 1930) 112

14. A Model of Production and Consumption in an Agricultural
    Economy with Non-agricultural Activities 128

LIST OF TABLES

1. Railway Mileage in China, 1894-1935 (Excluding Taiwan) . 11
2. Common Means of Transportation in China (Early 1930s) . 20
3. Average Agricultural Freight Rates of Major National Railways and Other Competitive Modes along Parallel Routes (c. 1920s) 22
4. Rail Transport Development in China: Selected Indexes (1916-26 Average = 100) 24
5. Estimation of Long-distance Agricultural Trade in Manchuria (1931-37 Average) 29
6. Railway Tonnage in Commercial Crops (1935): Selected Shares by Major Trunk Lines 46
7. The Share of Railways in the Interregional Transportation of Major Industrial Crops in China Proper (1935) . . . 48-49
8. Shares in Goods Arriving at Tientsin from Inland 54
9. Regression Relating Farm Output to Accessibility 70
10. Estimation of Equation (2'): Cobbs-Douglas Production Function Analysis of Chinese Agriculture, 1933 72
11. Regressions Relating Input Levels to Accessibility . . . 75
12. Regressions Relating Land-use and Labor Intensities to Accessibility 76
13. Factor-specific Output Elasticities with Respect to Accessibility 79
14. Regression Relating Labor Allocation Efficiency to Accessibility 82
15. Estimated Farm Output Response to Rail-induced Accessibility Changes 85
16. ANOVA Tests of Railway Effects on Farm Output and Yield 91
17. ANOVA Tests of Land-value Movements in Rail and Non-rail Localities 94
18. Testing for Differences in Commercial Crop Specialization in Railway and Non-railway Localities . 108
19. Regressions Relating Marketable Surplus to Accessibility 122
20. ANOVA Tests of Railway Effects on Marketable Surplus . . 123

21. Testing Allocative Efficiency of Labor
    in Chinese Agriculture   152

22. Testing Population Density Differences
    in Railway and Non-railway hsien   157

23. Testing Land-Tax Rate Difference
    in Rail and Non-rail hsien   160

24. Testing Land-Tax Rate Difference
    in Rail and Non-rail Sample Localities   163

ACKNOWLEDGEMENT

Many individuals have contributed to this study either by way of intellectual support or personal encouragement. Donald McCloskey first aroused my interest in the new economic history through his stimulating lectures in class. Donald Jones, who constantly displays a maximum of enthusiasm in cultivating the fruitful field of applying economic theory to the study of geographical phenomena, has been an indispensable source of intellectual guidance and friendly support. Norton Ginsburg taught me many things about China, both past and present, which, alas, I mistakenly thought I knew. George Tolley provided invaluable suggestions on the methodology and enlightened the interpretation of empirical results. Finally Chauncy Harris, with his wealth of knowledge, suggested many improvements without which this study would have been much less readable. To these individuals, teachers and friends they all are, I am gratefully indebted.

My thanks also go to Dr. Ramon Myers of the Hoover Institution at Stanford, for letting me research through the wealth of collections at the Hoover Institution and for his insightful comments on the study; to Professors Dwight Perkins of Harvard , Robert Dernberger of Michigan, and Anthony Tang of Vanderbilt, for both their encouragements and helpful comments on portions of the study; and to the participants at the Agricultural Economics Workshop of the University of Chicago, particularly Professors Theodore Schultz and D. Gale Johnson, for their stimulating discussion on the subject. Finally, I must thank my wife for her sacrifices and support during the years in which this study was slowly and painstakingly prepared, and to her this book is dedicated.

CHAPTER I

INTRODUCTION

Railways have long been regarded as among the most important, if not _the_ most important advance in technology, that affected entire economies in the nineteenth century. Several generations of economic historians have held unswervingly the view that the construction of railways marked an era of economic stimulation and rapid industrialization in many Western countries, particularly the United States.[1]

Until the pioneering works of Fogel and Fishlow,[2] however, the significance of railways' contribution to the process of national economic growth was never rigorously examined in quantitative terms. Both the novel methodology and unorthodox conclusions of Fogel and Fishlow, however, inspired a subsequent wealth of literature which added much to a better understanding of railways' historical role in the economic progress of many countries.[3] For nineteenth century America, the weight of new evidence on the effects of railways may be large enough to justify the view that railways were indispensable to the observed pace of economic growth, even if the resource savings[4] of the railways remained relatively small.[5] An important

---

[1] A brief review of the literature bearing out this persuasion can be found in Robert W. Fogel, Railroads and American Economic Growth (Baltimore: Johns Hopkins University Press, 1964), pp. 1-16; and in Patrick O'Brien, The New Economic History of the Railways (New York: St. Martin's Press, 1977), pp. 19-21.

[2] Fogel; and Albert Fishlow, American Railroads and the Transformation of the Ante-Bellum Economy (Cambridge, Mass.: Harvard University Press, 1965).

[3] For a summary survey of this literature see O'Brien.

[4] Resource, or social, savings is defined as the total economic benefit accruing to society as a result of the normal operation of its railway system over one year, assuming the volume and pattern of freight shipments remained unchanged.

[5] Robert W. Fogel, "Notes on the Social Savings Controversy," Journal of Economic History 39 (March 1979), p. 38. A recent study by Coatsworth on Mexico, however, came to the conclusion of railway indispensability in the growth process simply by demonstrating that the resource savings of railways was substantial; see John Coatsworth, "Indispensable Railroads in a Backward Economy: The Case of Mexico," Journal of Economic History 39 (December 1979):939-60.

dimension to the contribution of railways, in America as in a number of other countries, was the beneficial impact on agriculture, manifested in such causal processes as regional specialization, commercialization of production, and internal market integration.[1] The railway-induced increases in agricultural production not only improved agricultural incomes, which invariably constitute a very large component in the national income of backward economies, but it also tended to stimulate domestic industrialization through increases in the supply of food and industrial raw materials and in the demand for industrial products used as agricultural inputs.

The positive and potentially significant role of railways in the economic growth of early modern China is recognized, if implicitly, in a number of studies that emphasized the contributions of modern investments in China's early development efforts.[2] The actual impact of railways on the agriculture of pre-World War II China, however, has never been subjected to careful analytical investigations, though several authors have come up with some positive evidence in this regard in regional studies geared to some broader scopes of inquiry.[3] A few scholars, on the other hand, maintained that the existence of a sophisticated commercial system in pre-modern China was sufficient to downplay the potential role of modern rail transport in China's development and modernization efforts, and that both the traditional reliance on an extensive

---

[1] See, for example, Fishlow, especially chapter 5; John Hurd II, "Railways and the Expansion of Markets in India, 1861-1921," Explorations in Economic History 12 (September 1975):263-88; idem, "The Economic Impacts of Railways in India, 1853-1947," paper presented at the Economic History Workshop, University of Chicago, June 1976; Jacob Metzer, "Railroad Development and Market Integration: The Case of Tsarist Russia," Journal of Economic History 34 (September 1974): 529-49; and C. White, "The Impact of Russian Railway Construction on the Market for Grain in the 1860's and 1870's," in L. Symons and C. White, eds., Russian Transport: A Historical and Geographical Analysis (London: Bell, 1975).

[2] Hou Chi-ming, Foreign Investment and Economic Development in China, 1840-1937 (Cambridge, Mass.: Harvard University Press, 1965); Robert F. Dernberger, "The Role of the Foreigner in China's Economic Development, 1840-1949," in China's Modern History in Historical Perspective, ed. Dwight H. Perkins (Stanford: Stanford University Press, 1975), pp. 19-48; John K. Chang, Industrial Development in Pre-Communist China (Chicago: Aldine, 1969); and Shannon R. Brown, "The Transfer of Technology to China in the Nineteenth Century: The Role of Direct Foreign Investment," Journal of Economic History 39 (March 1979):181-98.

[3] For example, Ramon Myers, "The Commercialization of Agriculture in Modern China," in Economic Organization in Chinese Society, ed. W. E. Wilmott (Stanford: Stanford University Press, 1972), pp. 173-191; idem, The Chinese Peasant Economy: Agricultural Development in Hopei and Shantung, 1890-1949 (Cambridge, Mass.: Harvard University Press, 1970), esp. 192-93; John E. Schrecker, Imperialism and Chinese Nationalism: Germany in Shantung (Cambridge, Mass.: Harvard University Press, 1971); and Chou Shun-hsin, "Railway Development and Economic Growth in Manchuria," China Quarterly 45 (January/March 1971):57-84.

system of internal waterways and the spread of political turmoil in early modern China greatly reduced, if not totally removed, the imprint left by railways on the traditional economy.[1] It may be noted, however, that ex ante commercial sophistication need not evolve linearly into the industrial success characteristic of modern economic growth, though it provides certain necessary preconditions for industrialization.[2] Moreover, the potential contribution of modern mechanical transport, particularly railways, to the process of economic growth comprises not merely the expansion and integration of markets, but also the stimulative effects on the dissemination of modern technical and managerial know-how as well as on institutional reforms which are conducive to sustained economic growth.[3] Whether China's railways had been successful in these particular roles is, of course, a subject requiring elaborate empirical research. This study is an attempt to deal with a few of these issues.

The contention that water transport and internal disturbance in the political sphere diminished railways' beneficial impact on the traditional economy of China may be correct, but there is no a priori reason why it should be. The competitive efficiency of water transport is recognized in other countries where research has identified substantial contributions of railways to agricultural progress. It may also be noticed that some of the major trunk lines in prewar China ran in a north-south direction, in contrast to the mainly east-west flow-pattern of most major river routes.[4] The element of regional differentiation must therefore be highlighted, and any generalized conclusion based on the factor of water transport efficiency alone is unlikely to be satisfactory. Though there is no gainsaying the ruinous effects of political instability and civil warfare on the Chinese economy and on the efficiency of its railway system, these effects apparently were limited both in space and time.

---

[1] Rhoads Murphey, "The Treaty Ports and China's Modernization," in The Chinese City between Two Worlds, eds. Mark Elvin and William Skinner (Stanford: Stanford University Press, 1974), pp. 17-72; Albert Feuerwerker, The Chinese Economy, ca. 1870-1911 (Ann Arbor: Center for Chinese Studies, The University of Michigan, 1969); and Conrad Thomason, "Ambiguities of Modernization: The China Case," The China Geographer No. 5 (Fall 1976), pp. 13-27.

[2] Dwight Perkins, "Government as an Obstacle to Industrialization: The Case of Nineteenth-Century China," Journal of Economic History 27 (December 1967): 485-86.

[3] See the discussion in Moses Abramovitz, "The Economic Characteristics of Railroads and the Problem of Economic Development," Far Eastern Quarterly 14 (June 1955):169-78.

[4] One notable exception is the Grand Canal, but its commercial value as a through waterway had greatly diminished since the mid-1800s (see H. B. Morse, The Trade and Administration of China [London: Longmans, 1913], p. 322).

Besides, the negative influence of socio-political upheavals must be balanced by the positive impacts of domestic industrialization and urbanization on agriculture. Transformations in the urban and industrial sectors boosted farm output as development in China's rail transport proceeded. Also, further benefits on agriculture were conferred by an increasingly open economy.

In any case, whether railways had entailed any visible changes on China's traditional farm economy in the prewar decades remains largely an empirical question which cannot be answered without the systematic examination of empirical evidence. The potential role of rail transport in the economic development of early modern China, too, cannot be understood without investigating the causal mechanisms involved in such a relationship and the constraining factors that may have existed. It is the objective of this study to cast light on both these questions.

Structural changes in the agricultural economy in response to the development of railways may be expected to consist of spatial transformations in the pattern and composition of domestic agricultural trade, and in the pattern of agricultural production in regional and local economies as railways induced modifications in the market accessibility (economic distance to regional markets) of rural communities. The primary concern of this study is therefore with rural changes in relation to the economic interaction between the rural and the urban sectors via developing networks of rail transport. The methodology appropriate for such an inquiry differs from the graph-theoretic approach which has found common acceptance in the geographical literature of transportation and development.[1] The emphasis placed on market accessibility and the areal dimension of agricultural changes in this study serves to identify the micro-processes of economic responses of agricultural producers to the transmission of market incentives. Such a level of analysis is bypassed in the network analytic approach which recognizes a direct association between the structural complexity of transport networks

---

[1] See, for example, J. Nystuen and M. Dacey, "A Graph Theory Interpretation of Nodal Regions," Papers of the Regional Science Association 7 (1961):29-42; K. Kansky, Structure of Transport Networks, Research Papers, no. 84 (Chicago: University of Chicago Department of Geography, 1963). H. Gauthier, "Transportation and the Growth of the Sao Paulo Economy," Journal of Regional Science 8 (February 1968):77-94; R. Chorley and P. Haggett, Models in Geography (chapter 15) (London: Methuen, 1967); and E. J. Taaffe and H. Gauthier, Jr., Geography of Transportation (Englewood Cliffs, N.J.: Prentice-Hall, 1973), chapters 4 and 5. An application of the graphy-theoretic approach to the study of transportation development in China can be found in C. K. Leung, China: Railway Patterns and National Goals, Research Papers, no. 195 (Chicago: University of Chicago Department of Geography, 1980).

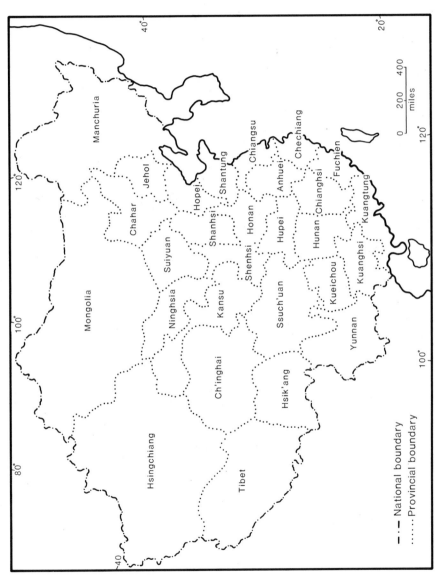

Fig. 1. Republican China: political regions of China proper (c. 1930)

and the level of regional development.[1] Emphasis on the structural efficiency of transport networks may, however, be misleading in the framing of development policies in low-income countries characterized by primitive network structures. Concern with network connectivity may dictate assignment of priority to investment in the line-haul (point-to-point, especially bewteen regional urban centers) development of transport systems, whereas agriculture would be benefited much more by the strengthening of structural linkages between urban and rural areas as well as the development of rural transport infrastructures at the local level.[2] The recognition of agricultural performance as a pivotal factor in the overall economic success of developing countries is an important reason to place emphasis on non-line-haul development of transport systems.[3] At early stages of economic growth, the constraint of capital scarcity would provide an additional reason for the allocation of available

---

[1] See, for example, Kansky; Gauthier, "Transportation and the Growth of the Sao Paulo Economy;" W. L. Garrison and D. F. Marble, The Structure of Transportation Networks (Washington, D.C.: U.S. Department of Commerce, Office of Technical Services, 1961); E. J. Taaffe, R. L. Morrill, and P. R. Gould, "Transport Expansion in Underdeveloped Countries: A Comparative Analysis," Geographical Review 53 (October 1963):503-29; and P. R. Gould, The Development of Transportation Pattern in China, Department of Geography Studies in Geography, no. 5 (Evanston, Ill.: Northwestern University, Department of Geography, 1960).

[2] Margaret Haswell, Tropical Farming Economics (London: Longmans, 1973), p. 115. It may be noted, however, that a well-conceived rural development plan may call for not only the improvement of transport facilities at the local level, but also the strengthening of organizational linkages, from marketing extension services to educational facilities, between the urban core and its rural hinterland; see M. L. Logan, "The Spatial System and Planning Strategies in Developing Countries," Geographical Review 62 (June 1972):289; Norton Ginsburg, "From Colonialism to National Development: Geographical Perspectives on Patterns and Policies," Annals of the Association of American Geographers 63 (March 1973): 16; Uma Lele, The Design of Rural Development (Baltimore: Johns Hopkins University Press, 1975), chapter 6; and Holland Hunter, "Chinese and Soviet Transport for Agriculture," (Pittsburgh: University of Pittsburgh, University Center for International Studies, 1974) (mimeographed).

[3] During the past two decades, the role of agriculture in the economic growth of low-income countries has found increasing emphasis in the development literature. The consensus now among development specialists appears to be that an effective economic development strategy, particularly during the early stages of economic growth, depends critically on productivity growth in agriculture. See, for example, Bruce Johnston and John Mellor, "The Role of Agriculture in Economic Development," American Economic Review 51 (September 1961):566-93; H. J. Habakkuk, "Historical Experience of Economic Development," in Problems in Economic Development, ed. E. A. G. Robinson (London: Macmillan, 1965), chapter 6; Bruce Johnston, "Agriculture and Structural Transformation in Developing Countries: A Survey of Research," Journal of Economic Literature 8 (June 1970):369-404; T. W. Schultz, Economic Growth and Agriculture (New York: McGraw Hill, 1968); H. M. Southworth and B. F. Johnston, eds., Agricultural Development and Economic Growth (Ithaca: Cornell University Press, 1969); and Y. Hayami and V. W. Ruttan, Agricultural Development: An International Perspective (Baltimore: Johns Hopkins University Press, 1971).

resources away from grandiose transport projects that are often needed to increase network complexity but may not be justified because of lower returns to society than an aggregate of well-conceived projects of much smaller scales.

The scope of this study is restricted to China proper both because of data requirements and of the fact that Manchuria was largely under the political and economic domination of foreign powers, notably Russia and Japan, during the time period under consideration. The time span studied covers the six decades since rail transport was first introduced into the country in 1875. The choice of 1935 as the year of termination for this study is not entirely arbitrary, for the year 1935 marked the peak of recovery of the economy, in particular the agricultural sector, from the post-1930 general depression, and the national railway system from the disruption entailed by a decade's internal warfare, before full-scale war with Japan broke out in 1937.

The historical pattern of railway development is briefly surveyed in the following chapter, with special attention given to the effects of changes in the political and economic environments. The development of rail transport is also discussed with respect to efficiency aspects, both in terms of its potential competitiveness with other traditional modes of transport and in terms of its actual performance, as reflected in operational efficiency indicators, during the prewar decades. Apart from periods of serious political disturbance and regions well-served by water transport, these analyses indicate that Chinese railways appeared to be a promising force for the transformation of the traditional agrarian economy.

Chapter III takes on the important task of tracing the relationship between the growth and diversification of interregional agricultural trade in China proper and the expansion of railway transportation. This relationship is exposed through (1) spatial changes in the pattern of interregional trade, (2) the quantitative share of railborne freight traffic in total long-distance agricultural trade, and (3) the synchronic growth of export trade in major treaty ports and railway extensions. The findings of this chapter provide a macroscopic view of the effects of railways on agricultural transformation. The underlying micro-mechanisms of this transformation process, however, are closely analyzed in the next chapter.

The empirical relationship between agricultural productivity growth and railway construction is statistically identified in chapter IV. This is done via two different approaches. The first approach makes extensive use of a body of primary cross-section data on agricultural production at the farm level to verify that railway construction improved market accessibility of farmers and thereby stimulated farm output by encouraging the added employment of scarce resources and their more efficient utilization in the production process. The second approach draws on a body of time-series data on farmland values and directly identifies the positive correlation of the growth in land values, and thus the growth in agricultural productivity, with the construction of railway lines. The exogeneity question of railways in the growth of agricultural productivity is also treated, separately in the appendices, by examining historical evidence, though both the time-series analysis of farmland values and pertinent evidence examined in the following chapter provide partial answers to it.

Chapter V is devoted to investigating the impact of railways on agricultural commercialization. Statistical analyses of regional specialization patterns in selected cash crops help bear out the perceptible imprints of railways on agricultural commercialization. This finding is reinforced by the direct establishment of a strong empirical relationship between market accessibility and marketable farm surplus. The institutional factors that promoted specialization along railway lines are also discussed, which apparently contributes to a better understanding of the role of railways in rural development.

Chapter VI has two objectives. First it seeks to elucidate the effect of trade expansion on rural economic development by means of a model which highlights the functional relationship between agricultural and non-agricultural production activities in the rural sector. The relationships between urban-industrial growth, rail transport expansion, and rural economic development can therefore be exposed in a general-equilibrium framework. The second objective of this chapter is to give a thorough examination of the sources of constraints on the beneficial impact of railways on the peasant economy of prewar China. Through this analysis it may become clear why Chinese railways failed to perform more effectively in stimulating the rural economy. Finally, the policy implications of these findings are summarized and discussed in the closing chapter.

CHAPTER II

THE DEVELOPMENT OF RAIL TRANSPORT IN CHINA

Railway History to 1935: An Overview

The early course of railway building in China was set by the interplay of an aggressive mercantile interest of foreign powers in the country, as a market for industrial exports and a supplier of rich natural resources, and a nationalistic movement which was thrust into being by the humiliation of undeterred foreign encroachment. Ironically, foreign capital, both financial and physical, was singularly responsible for the physical existence of a national railway system in China, one which in the eyes of many progressive late Ch'ing officials and intellectuals would be instrumental in modernizing the economy and finally delivering the country from the menace of foreign incursions.[1] This dependency on imported capital rendered domestic railway development vulnerable to upheavals in international relations and in the economic conditions of capital-exporting countries. In effect, it proved to be a hindrance to the development of railway transportation in China, with the exception of Manchuria, in the entire pre-Communist era.[2]

---

[1] The strategic and defense functions of railway construction were emphasized by officials who initiated and steered the course of the self-strengthening" movement in the last decades of the Ch'ing dynasty (for a brief history of the movement, see John K. Fairbank et al., East Asia: The Modern Transformation [Boston: Houghton Mifflin Co., 1965], chapter 5). A crude scheme of a national system of railways was incorporated by Liu Ming-chuan and Li Hung-chang in their court memorials of 1880 (see Pi Yu-cheng, ed., Chung-kuo chin-tai t'ieh-lu-shih tse-liao [Documents on the history of Chinese railways of modern times] [Peking: Chung-hua Book Co., 1963], pp. 87-91).

[2] The ostensible causes of China's financial debility throughout her late Imperial and early Republican years originated in the huge sums of indemnity payments due from her as the vanquished in Ch'ing's skirmishes with foreign powers. (One of the largest debts China has ever contracted was an aftermath of the Boxer Uprising in 1900, a frenzied anti-foreign movement of the populace with the patronage of the Ch'ing court.) Such indemnity obligations constituted about 43% of the total revenue of the Imperial treasury in the early 1900's and, as late as 1931, foreign debt payments (interest and amortization) still amounted to 36.3% of the actual cash receipts of the National Government even when excluding the Boxer Indemnity (Hou, p. 45). Inevitably, social overhead capital investments in early modern China were largely financed through foreign loans, which accounted for as much as 80% of the country's railway investment in the 1920's (E-tu Zen Sun, "The Pattern of Railway Development in China," Far Eastern Quarterly 14 [February 1955]:188). The insecurity of such dependence on imported capital first manifested itself with the onset of the First World War, which inaugurated an era

Chinese railway history began in 1875, when a ten-mile track connecting Shanghai and Woosung was built by the British Jardine, Matheson & Co. without prior approval by Chinese authorities. Though apparently an immediate financial success, the line was soon purchased by the Chinese Government and demolished, an outcome largely attributable to Chinese apprehension about foreign control of a means of modern transport that was of extreme strategic value.[1] Intellectual conservatism and official inertia effectively impeded the materialization of several railway construction proposals in the ensuing decade,[2] with the notable exception of the six-mile Kaiping Tramway for coal conveyance at Tangshan, which was completed in 1879 and was ultimately extended to become the Peking-Mukden Railway.[3]

---

of sluggish development in China's railway industry.
 It is easy to blame foreign economic imperialism for China's apparent underdevelopment, but any discerning student of Chinese history would doubt China's ability to achieve financial independence in her modernization efforts even in the absence of large indemnity payments. The inability of the Manchu Government to institute reforms in the centuries-old land-tax basis (Wang Yeh-chien, Land Taxation in Imperial China, 1750-1911 [Cambridge, Mass.: Harvard University Press, 1973], p. 131); a venal court; the lack of central control over the absolutism of officials; and the conspicuous consumption of the ruling class all laid unmistakably in the roots of the country's financial woes (see S. R. Wagel, Finance in China [Shanghai: North China Daily News & Herald, 1914], pp. 1-50). In the Republican period, it was internal warfare, political disintegration, heavy taxes, and corruption that bedeviled the state of economic affairs. Exogenous economic assistance in the development of backward economies is also necessitated by the scarce investible capital resources with which these economies were to begin, but the benefits of such external aid must be sought in the long-term educational and stimulative effects of knowledge and technological transfer. In these respects the railway enterprise of early modern China had largely failed (Abramovitz, p. 176), a result that may be traced back to governmental inaction in the general spheres of economic and institutional reform. (For a more penetrating discussion of some of these points see Perkins, "Government as Obstacle.") One is therefore not surprised to discover the plight of many early Chinese railways underlaid by such factors as financial mismanagement, disintegration of control and planning, repressive duties on trade and marketing, and stagnant development of complementary infrastructures and industries. The assertion that excessive financial charges on foreign railway loans were responsible for the financial failures of Chinese railways is therefore based on superficial observations. (Such an argument is maintained by, for example, Frederick V. de Fellner, Communications in the Far East [London: P. S. King & Son, 1934], pp. 118-119.)

 [1]Percy Kent, Railway Enterprise in China (London: Edward Arnold, 1907), pp. 9-15; and Li Kuo-ch'i, Chung-kuo tsou-ch'i ti t'ieh-lu chiung-yung (History of the early Chinese railway development)(Taipai: Chung-hua, 1961), pp. 37-45. The alleged cause of the dissolution of the enterprise was its accidental killing of a Chinese national, though the accident itself was suspected of being contrived (Kent, p. 13).

 [2]Lee En-han, China's Quest for Railway Autonomy (Singapore: Singapore University Press, 1977), p. 11.

 [3]Wang Ch'in-yü, Chin-tai chung-kuo ti tao-lu chien-sheh (Road and railway construction in modern China)(Hong Kong: Lung-man, 1969), p. 70; also Hsieh Pin, Chung-kuo t'ieh-lu shih (A history of Chinese railways)(Shanghai:Chung-hua, 1929), pp. 330-35.

Until the outbreak of the Sino-Japanese War in 1894, no more than 195 miles of railway tracks were built on the Chinese mainland.[1]

China's ignominious defeat in the Sino-Japanese War of 1894-95 inaugurated an epoch of vigorous railway development not to be repeated in China's pre-Communist history. Close to 6,000 miles of tracks were completed between 1894 and 1914, though only 53% of these were in China Proper (table 1). The spree of railway

TABLE 1

RAILWAY MILEAGE IN CHINA: 1894-1935
(EXCLUDING TAIWAN)

Year	China Proper	Manchuria	Total	Average Annual Growth Rate for Period ending with Year
1894	195	---	195	---
1914	3,325	2,727	6,052	18.74%
1935	5,485	4,288	9,773	2.43%

SOURCES: 1894 figure from Lee, p. 13; 1914 and 1935 figures calculated from data in Wang, pp. 73-78, 84-93.

construction was ushered in by the so-called "Battles of Concessions,"[2] in which was climaxed the imperialist rivalry for China's territorial rights and commercial sovereignty within the respective "spheres of influence" of the foreign powers. As a consequence, most of the railway lines falling in the category of direct foreign investment and control were constructed during this period,[3] which can be said to have lasted until the Russo-Japanese War of 1904. The war patently testified to the emergence of Japan as a self-made world power. The psychological repercussions of this fact on the Chinese public were such as to invigorate a nationalistic movement characterized by deep-seated rancor against foreign infringement.

One aspect of this nationalistic movement was reflected in the surge of investment by merchant-gentry groups in railway enterprises.

---

[1] Lee, p. 13, table 1. There were also 54 miles of railway in Taiwan.

[2] The phrase was coined by Britain's Lord Salisbury to describe the new efforts on behalf of introduction of railways into China (C. A. Middleton Smith, The British in China [London: Constable, 1920], p. 157).

[3] The more important lines built with direct foreign investment included the Germans' Chinan-Ch'ingtao line in Shantung, the French Tongking-K'unming line in Yunnan, and the Russian's Manchuli-Suifenho and Harbin-Port Arthur system in Manchuria. The Harbin-Port Arthur line was taken over by the Japanese after the Russo-Japanese War to become the South Manchurian Railway.

As a result privately financed Chinese railways came to share about 5% of the total track mileage in the country by 1912.[1] Sheer patriotism, however, was hardly the only factor that sparked the boom of private railway building. The effect of institutional reforms, notably in the establishment of a Ministry of Commerce in 1903 and its subsequent efforts in the consolidation of commercial laws and regulations, together with an official attitude to promoting indigenous investment in railways, was surely one of strong stimulation.[2] Yet the greatest impetus to domestic private investment in the highly risky railway business was the demonstrative success of earlier-built, foreign-financed enterprises, which by the mid-1900s had begun to display handsome returns probably unmatched by most alternative investment opportunities.[3] The channelling of domestic savings to modern productive investments such as railway construction had important implications for China's early economic development, and foreign investment in China certainly played a role in fostering such changes.

The burgeoning nationalistic movement of the 1900s had, of course, a broader basis than just the merchant-gentry class. The cumulative experience of defeats and Imperial Japan's economic success had dissipated much of the resistance to modernization schemes from the conservative bloc of China's ruling class, even if the implementation of such schemes necessarily meant requisitions for foreign capital. A government eager for modernization and an equally eager line of foreign syndicates thus worked to create, in just over a decade, the main skeleton of modern China's railway system (figure 2). Most of the major trunk lines in China's prewar railway network were either laid down or already under construction by the time international tension in Europe developed into a full scale war in 1914.

---

[1] Grover Clark, Economic Rivalries in China (New Haven: Yale University Press, 1932), figure facing p. 21.

[2] For a discussion of these institutional reforms and their relation to the formation of the nationalistic movement, see Lee, pp. 9-44.

[3] The Imperial Railways of North China (I.R.N.C.)(later called the Peking-Mukden railway), for example, had a 19% rate of return to invested capital in 1905 (based on profit figures quoted in Kuo-feng-pao [Journal of public opinion] 2 [1906?]:11), just one year after its completion. The Japanese-managed South Manchurian Railway, on the other hand, reported a remarkable 40.3% return rate in 1907 (Hou, p. 114). Even the barely finished Peking-Hankow line registered a net profit of some 2.4 million taels in 1904, or close to a 7% rate of return on investment cost (Lee, p. 35). These profit rates undoubtedly compared favorably with the average return on the traditional absorbant of domestic private savings, land, which probably amounted to no more than 5% in net return and was probably declining in profitability as an investment asset; see Dwight Perkins, Agricultural Development in China, 1368-1968 (Chicago: Aldine, 1969), p. 95.

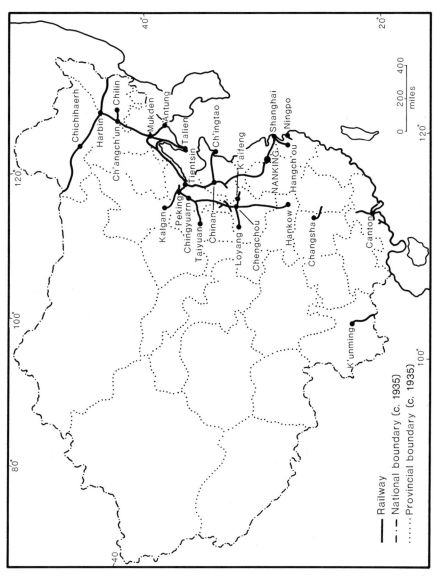

Fig. 2. Railways in China, 1914 (compiled from information in Wang Chin-yü)

The outbreak of the First World War was followed by two decades of slow progress in Chinese railway development. From 1914 to 1935, barely over 2,000 miles of tracks were added to the railway network in China proper, whereas Manchuria gained more than 1,500 miles of railway during the same period (see table 1). Most of the new additions in China proper represented only extensions of the existing network or completion of projects under progress before 1914 (figure 3). Nevertheless the completion of such trunk lines as the Lung-Hai (Lienyungkan to Hsian), Peking-Paot'ou, and Hankow-Canton railways[1] within this period undeniably contributed substantially to the maturity of the country's prewar railway system.

A number of events had conspired to bring about this slackening phase of railway extension. In spite of all the virtues of a progressive railway policy adopted by the Republican government after its establishment in 1911,[2] railway building quickly suffered from the cessation of Western capital inflow upon the start of the First World War. Dearth of capital throughout the war years due to the termination of Western capital supply was directly responsible for the abandonment of some major railway projects and delayed the realization of others.[3] However, one must not ascribe presumptuously the diminution of foreign capital inflow to economic strains confronting the war economies during and even after the War. China's reputation in the international money market was blemished when she left some outstanding loan payments in default since 1917. This apparently discouraged further Western investment in China's

---

[1] The Hankow-Canton line was finished in early 1936 (Wang Ch'in-yü, p. 90).

[2] The new policy led to the government's arrangement with foreign banking institutions for a score of railway loans in 1913 and the appointment of a commission to the unification of railway accounts and statistics; see H. Stringer, The Chinese Railway System (Tientsin: Tientsin Press Ltd., 1925), pp. 15-16.

[3] An important scheme that was so affected was the Yangtze scheme, which included the projected construction of the Hankow-Ssuch'uan line (Maritime Customs, Decennial Reports, 1921-31 [Shanghia, 1932],1:556). It must be noted, however, that although the initial obstacle to the undertaking of the Hankow-Ssuch'uan project was the apparent lack of capital, it was probably the development in river transport that delayed the project even to this date. Before steamers were successfully introduced to the scene, the extremely hazardous navigation conditions in the Yangtze Gorges posed well-nigh insurmountable barriers to the sizable growth of riverine trade between the lower Yangtze basins and Ssuch'uan, thus giving rise to the prospect of rail transport development. Export trade at Ssuch'uan's major Yangtze port, Ch'ungch'ing, however, increased by leaps and bounds once steamer service was developed in the early 1920s (Chu Chien-pang, Yangtze-chiang hang-yeh [River Transport on the Yangtze River][Shanghai: Commercial Press, 1937], p. 86). Expectedly, steamshipping was a highly profitable business on this route and threatened even the survival of traditional river junks (Cheng Ming-ju, The Influence of Communications on the Economic Future of China [London: G. Routledge & Sons, 1930], p. 29).

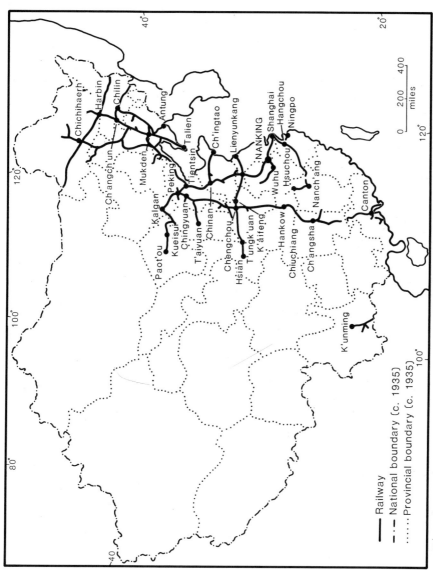

Fig. 3. Railways in China, 1935 (compiled from information in Wang Chin-yü)

railways.[1] An even more conspicuous deterrent to foreign investment was perhaps the escalating political instability in the country since 1916, since economic uncertainty added to the risk and curtailed the expected return of long-term investments. These factors obviously accounted for the transaction of certain high-interest, short-term loans in the early 1920s. The financial burden of these debts in turn exacerbated the economic predicament that befell Chinese railways in the 1920s and 1930s.[2]

The ebbing of Western capital was contrasted, however, by a marked intensification of Japanese investment in China.[3] The Japanese, of course, had a long-term interest in the country which was not merely economic, as was manifested by the Mukden incident in 1931 and subsequent open military invasions. By 1931, Japanese direct investment in China had amounted to more than 35% of total foreign investment in China, as compared with only 13.6% in 1914.[4] Nearly 37% of the railway loans contracted by the Chinese during this 15-year period was of Japanese origin. This was nearly twice as large as that from the British, the predominant source of China's railway capital before 1914 and now second in importance;[5] but the shortfall of imported capital from Western countries did not merely create an opportunity for the expansion of Japanese economic interests, which was then still largely confined to Manchuria. The circumstance was swiftly seized upon by Chinese entrepreneurs and the country's young but strengthening group of capitalists as a good opportunity to curtail foreign dominance in Chinese railways as well as to spend, in a productive manner, the accumulation of private capital.[6] As a result, both the state and private banks began to play an increasingly active role in railway finance after 1920. Much official encouragement was also given to such activities after the

---

[1] Sun, pp. 189-90.

[2] Financial charges of those loans were reported to be as high as 9.5% p.a. with an amortization period of only one year (Sun, p. 188). These severe terms were in stark contrast with the much lower interest and long-term nature of earlier loans.

[3] Japan's ascendancy over other Western powers in the arena of economic rivalries in China was set off by the presentation of the Twenty-One Demands in 1915, in which exclusive privileges were demanded for Japanese subjects in Manchuria and railway rights not only in Shantung and Manchuria but in the Yangtze valley as well; see Stringer, p. 17.

[4] Hou, p. 17.

[5] Wang Ch'in-yü, pp. 81-83.

[6] Sun, p. 189.

unification under the National Government in 1928.¹ Of all the track mileage constructed between 1914 and 1935, those financed by purely Chinese concerns accounted for about 78%, as compared with only 23% before 1914.²

The languid development of Chinese railways in the post-1914 era, however, belied the pace of economic progress in the country's other sectors. The economic boom touched off by the First World War was, perhaps fortuitously, a great blessing to China's inchoate industrial base and her indolent peasant economy whose capacity was strained by population pressure.³ China's farm economy, in particular, was well stimulated by more than a decade of rising agricultural prices and, capitalizing on an expanding modern transport system, it became increasingly drawn into the national and even international market.⁴ It was, therefore, unfortunate that China, while being afforded a chance propitious for sustained economic growth, was plagued by internal warfare in the mid and late 1920s. Military disturbances arising from clashes among regional warlords was practically eliminated after the establishment of the National Government in 1928, but then military campaigns against the Communists began to tax the country's economic resources. Widespread depression of the late 1920s and early 1930s and rising silver prices on which China's currency system was based until 1936 both adversely affected the country's economic conditions.⁵ As far as rail transport performance is concerned, the havoc wrought by the armed strife among warlord cliques was immediate and profound. Needless to say, agriculture fell prey to the sequence of political disturbance too, both directly as a result of systematic pillaging and over-taxation, and indirectly because of the disruption of trade and deterioration in railway efficiency.

---

¹Ibid.

²In 1914, about 2,200 miles of railway track in a total of 6,000 miles in the entire country could be classified as entirely Chinese, i.e., financed by Chinese capital, either from governmental or private sources. By 1935, more than 50% of the then existing 9,700 miles of railways fell in the same category. About 2,900 miles of the tracks added between 1914 and 1935 were built with Chinese capital, while only about 800 miles were financed by non-Chinese resources. These figures are calculated from data in Wang, pp. 73-89.

³For an analysis of China's war-stimulated industrialization see Cheng Yu-kwei, Foreign Trade and Industrial Development of China (Washington, D.C.: Washington University Press, 1956), pp. 27-36.

⁴Historical evidence on the impact of rising prices on agricultural production will be examined in chapter 5 below. Railways' role in the expansion of interregional trade will be discussed in the following chapter.

⁵Cheng, pp. 67-71.

Rail Transport Development: Efficiency Aspects

In spite of the prevalence of an inimical attitude toward Westerners, the Chinese literati were well disposed, even at an early time, to regard highly the value of Western technology. This was, of course, what generated the momentum of the "self-strengthening" movement in the last decades of the Ch'ing dynasty. Yet in no form was the introduction of Western technology in China so popularly and instantly appreciated than the "iron horse," even if its presence necessarily provoked bitter sentiment on the part of local inhabitants because of its interference with the geomantic spirit of wind and water (feng-shui) and its destruction to the sacrosanct graveyards. The success of some early railway enterprises, as referred to above, testified to the anticipated demand for this modern form of overland transport in traditional China.

This early experience of prosperity was, moreover, hardly an ephemeral one. The prewar Chinese railway system fared remarkably well as a public enterprise. Civil wars of the late 1920s and economic depression of the early 1930s were debilitating to the earning capacity of many trunk lines, but the railway system as a whole still managed to earn an average annual return of 7.4% on paid-in capital, and 3.3% after meeting interest obligations, during the two decades from 1916 to 1935-36.[1] Nor was strenuous competition from river transport a particularly severe threat to railway earnings. The Shanghai-Nanking line, for example, was able to pay 9-10% on invested capital in the 1910s, despite the fact that it served a region crisscrossed by rivers and canals.[2] The Government, in fact, creamed off an annual average of C$12.7 million,[3] which was some 28% of the average annual net revenue of the railways between 1916 and

---

[1] Hou, p. 41. The ratio of net operating revenue to the cost of road and equipment averaged 8.32% between 1916 and 1925, as compared with 5.6% for 1926-1935/36. The ratio of net profit to cost of road and equipment, representing the after-interest rate of return, was 6.3% for 1916-25 and a mere 1.6% for 1926-1935/36 (ibid.). This indicated that foreign railway loans to the Chinese Government were at least self-liquidating. It is also interesting to compare the above figures with America's early railroad earnings. For trunk lines in the American Mid-West, gross return rate averaged 5.6% in 1849, 7.2% in 1855-56, and only 3.7% in 1859 (Fishlow, p. 178).

[2] C. A. Middleton Smith, The British in China (London: Constable, 1920), p. 167.

[3] C$ stands, hereafter, for Chinese Yuan, the official currency unit of the National Government. C$1.00 = U.S.$0.363 in 1935 and U.S.$0.299 in 1930 (see National Tariff Commission, Prices and Price Indexes in Shanghai [Shanghai, 1937], 13, no. 1:20).

1935.[1] This, at least in part, was responsible for the default of interest payments on the railway loans. While burdensome interest obligations might represent risk premiums to lenders, the sizable remittances to government appeared as an unwarranted drain on railway earnings. This was also a major cause of the under-maintenance and slow productivity growth of Chinese railways throughout the prewar decades.

The demonstrated viability of Chinese railways was certainly not a unique experience, for comparable, if not more spectacular, success often characterized railway enterprises in other countries. However, the advent of rail transport did create a unique opportunity for a country whose better alternative to an imminent future of deterioration in living standards lay clearly in successful efforts at industrialization and agricultural progress. No matter what the extent to which railways in China actually promoted the achievement of these ends, economic transformation into a sustained growth process for the nation would indeed be very much handicapped in the absence of simultaneous development in transport technology.

The touted economic advantages embodied in the technological superiority of railways over traditional means of transportation constituted a common denominator in the economic importance and financial success of railways built even in grossly dissimilar environments. Table 2 summarizes some of these comparative characteristics of various common means of freight transport in prewar China. Although the apparent supremacy of railways in speed and carrying capacity is generally acceptable, the comparison in terms of unit transfer cost admittedly glossed over regional differences which could be substantial. A more meaningful comparison is in terms of average unit costs on shipment routes parallel to and in proximity of railway lines. The results of this exercise are displayed in table 3, which reveals the possible unit savings in direct transfer cost by railways on specific routes of transportation.[2] It may be noticed that such savings were generally positive for trunk lines in

---

[1] Hou, p. 40.

[2] The regression estimates of average freight rates on major railway lines underestimated short-distance rates but overestimated long-distance rates (due to the tapering rate structure). Since most agricultural shipments on railways were long-distance, overestimation is the predominant case (thus underestimation of possible savings by railways). Note also that in North China (see n. 2 below) seasonal restriction on river navigability is an important factor in the low competitiveness of river transport (about 230 days were navigable in major waterways of North China [Chiang Wu-chin, "Min-chuan chih yün-hsiao cheng-pen," (Cost of river junk transport) Chiao-t'ung tsa-chih 3 (January 1935):20]). The present calculations, however, fail to reflect this cost.

TABLE 2

COMMON MEANS OF TRANSPORTATION IN CHINA
(EARLY 1930s)

	Capacity (kg.)	Speed (km.)	Cost (C$/ton-km.)
Railway	15,000-40,000[a]	32-45/hr.	0.02[b]
Motor Vehicle	770-3,000	40-60/hr.	0.32[c]
Cart[d]	300-1,500	40-50/day	0.05-0.17
Wheelbarrow[e]	180-300	40-50/day	0.10-0.14
Donkey, Mule, and Horse	100-150	40-50/day	0.20-0.35
Camel	120-200	40-50/day	0.10-0.20
Ricksha	180-240	40-50/day	0.20-0.35
Human Porterage	50-70	40-50/day	0.20-0.35
Native Junk	1,000-20,000	48-96/day	0.20-0.13
Steamship	5,000-3,000,000	6-25/hr.	0.02-0.16

SOURCE: Bureau of Roads, <u>Chung-kuo ti kung-lu</u> (Highways in China)(Nanking: National Economic Council, 1936), table 15.

[a] Capacity of one freight coach.

[b] Average rate charged for all agricultural goods conveyed in 1935 by the National Railroads (Yang Hsiang-nien, <u>T'ieh-lu ching-chi yü t'sai-cheng</u> [Railway Economics and Finance][Shanghai: Commercial Press, 1944], p. 109).

[c] Average rate for second and third class goods; these goods correspond more closely to the classification of general agricultural commodities in railway regulations.

[d] Vehicles drawn by one to four draft animals.

[e] Vehicles driven by one or two men. Higher capacity and lower cost figures pertained to two-men-drawn wheelbarrows.

North China,[1] particularly in regions where water transport was not well developed or simply impossible (e.g., along the Peking-Paot'ou and Peking-Hankow lines). In South and Central China, where water transport was traditionally an important form of internal communication, much of the advantage enjoyed by railways in direct shipment cost disappeared. The lower Yangtze delta area in particular was well served by old-fashioned river junks and, by the early 1920s, by ocean-going steamers. The competitiveness of water transport is well reflected in the comparative freight charges along shipment

---

[1] The terms North and South China in this study correspond loosely to the Wheat and Rice Region, respectively, as defined in Buck (J. L. Buck, <u>Land Utilization in China</u> [Nanking: Nanking University Press, 1937]). The approximate geographical boundary of the two regions can be found in figure 7, p. 68.

routes parallel to the Nanking-Shanghai and Shanghai-Hangchou-Ningpo railways (nos. 7 and 8 in table 3).

Direct cost of shipment represents only one component in the total cost of transportation, however. In other countries a large part of the resource savings generated by railways came not from lower freight charges, but rather from such indirect savings as higher speed, greater dependability, increased regularity, and more flexibility (in routing).[1] Direct estimation of savings from these sources is not possible for Chinese railways because of data insufficiency, but it may be assumed that, as in other countries, these embodied benefits of rail transport added to the competitiveness of railways. Reduction in such risk elements as "shrinkage" (cargo loss en route), the institutionalization of freight insurance, and the possibility of adjusting physical flows to market conditions are especially important innovations associated with rail transport which appealed to dealers of more valuable commodities. The observed viability of trunk lines faced with strenuous water competition was almost certainly rooted in the exercise of these less visible advantages of rail transport.

The comparison of freight costs provides a picture of the potential competitiveness of Chinese railways as a long-distance carrier of agricultural freight. It tells little, however, of the actual performance of the railways in the carriage of agricultural commodities, which of course is affected both by the relative competitiveness in terms of cost savings and the operational efficiency of the railway system. Apparently it is this actual performance that is of importance in understanding the effect of railway development on the agricultural sector. A glimpse of this performance can be obtained from table 4, column (C), which registered five-year interval variations in agricultural and livestock traffic on Chinese National Railways between 1916 and 1935. For a better understanding of the general efficiency of the railway system during the period 1905-1935, several related indexes with more complete data coverage are also presented in table 4 (cols. A, B, D). In general, these four indexes portrayed a consistent upward trend in the performance of the railway system between the years 1905 and 1925. As has been alluded to, this twenty-year period was marked by first the rapid gain in railway mileage and then increased robustness in the economy, both factors that contributed to railway efficiency.

The drastic deterioration in rail transport efficiency during the peak civil war years of late 1920s was clearly reflected in the

---

[1] For an attempt to quantify some of these cost components see Fogel.

TABLE 3

AVERAGE AGRICULTURAL FREIGHT RATES OF MAJOR NATIONAL RAILWAYS AND OTHER COMPETITIVE MODES ALONG PARALLEL ROUTES (c. 1920s)

	Average Cost (C$/ton-km.)			
	Railway[a]		Competitive Modes(s)	
Railway Routes	Non-Carload	Carload	Mode I	Mode II
1. Tientsin-P'uk'ou	0.0200	0.0132	Junk: 0.0270[b]	Steamer: 0.0280[c]
2. Peking-Mukden	0.0383	0.0254	Steamer: 0.0280[d]	...
3. Peking-Hankow	0.0520	0.0390	Cart: 0.0746-0.162[e]	...
4. Peking-Paot'ou	0.0701	0.0438	Camel: 0.1243[f]	Cart: 0.0559-0.0810[g]
5. Ch'ingtao-Chinan	0.0252	0.0168	Cart: 0.1056-0.1305[h]	...
6. Lienyunkang-T'ungk'uan (Lung-Hai)	0.0217	0.0190	Junk: 0.1234[i]	...
7. Nanking-Shanghai	0.0385[j]	0.0116[j]	Junk: 0.0124-0.0497[k]	Steamer: 0.0373-0.0487[l]
8. Shanghai-Hangchou-Ningpo	0.0428[m]	0.0255[m]	Junk: 0.0186-0.0336[n]	...

[a] The average freight rates charged by major national railways were estimated by regressing total freight costs on respective trip distances as given in Kao Lu-ming, T'ieh-lu yün-chia chi yen-chiu (A study on railway freight charges) (n.p., n.d.), pp. 159-62. Agricultural freight classification on each line follows those given in Ministry of Communications, Chiao-t'ung shih, Lu-cheng-p'ien (Section on railways, history of communications in China)(Nanking, 1930), 18 volumes. Most agricultural commodities fell in the 3rd, 4th, and 5th classes, but the class comprising major farm staples, e.g., cereals and common cash crops, was used. The rates used by Kao were the basic rates which became effective in 1921. A general reduction in freight rates, however, was decreed in 1932 (see Yu Yen, Tsui-chin san-nien t'ieh-lu chien-ti-yün-chia shu-lueh [A brief account of the reduction in railway freight rates in the past three years][Nanking: Ministry of Communications, 1935]). These rates, however, are not available for general comparison purposes.

[b] This is the average rate charged by junks plying on the Grand Canal between Nanking and Tientsin. Three separate sectional rates may be identified. For the northern section of the Canal between Chinan (Shantung) and Tientsin (Hopei), the long-distance junk rate recorded by Buck for Chinan was used (J. L. Buck, Land Utilization in China, Statistical Volume [hereafter LUSV][Chicago: University of Chicago Press, 1937], p. 347). For the southern section from Huaiyin (Chiangsu) was used (ibid.). These two rates were C$0.04 and C$0.02/ton-mi., respectively. For the middle section between Huaiyin and Chinan, a higher rate of C$0.05/ton-mi. was assumed, since this section of the canal was badly silted and little used by the 1900s (see H. B. Morse, The Trade and Administration of China [London: Longmans, 1913], p. 322; see also Stringer, p. 116). The distance weighted average

of these three rates is C$0.0435 per ton-mile, or C$0.027/ton-km., which has been used as the competitive rate charged by canal junks.

^{c}This represents the average rate agreed on by major steamship companies for shipping tea from Shanghai to Tientsin in the late 1920s (Shanghai te-pieh-shih she-hui-chu [Shanghai Special Municipality Bureau of Social Affairs], Shanghai chi kung-yeh [The industries of Shanghai][Shanghai: Chung-hua Book Co., 1930], Section II, p. 47.

^{d}Since relevant steamship rates cannot be obtained, the Shanghai-Antung steamer rate for shipping tea is used as a proxy for the Yingk'ou-Tientsin rate. This rate is in all likelihood an underestimate of the actual rate because of the much longer distance involved.

^{e}The only route that ran parallel to the Peking-Hankow line was the old trade and postal route that leads from Hankow (Hupei) to Peking (Hopei) via Sinyang, K'aifeng (Honan), and Chengting (Hopei)(see Cheng, p. 59). Since this is a land route, the average long-distance rate for carts in the provinces of Hupei, Honan, and Hopei is used. The lower figure of C$0.0746/ton-km. was the average rate in Hupei and the higher figure of C$0.162/ton-km. was the average rate in the Winter-Wheat-Kaoliang Crop Region (Buck, LUSV, p. 347).

^{f}This is the average unit cost of camel transport from Lanchou (Kansu) to Paot'ou (Suiyuan) in the 1920s and was taken from Wang Ching-wang, tr., Chung-kuo hsi-pei chih ching-chi chuang-k'uang (The economic condition of China's Northwest) (Shanghai: Commercial Press, 1933), p. 57. Reports in the early 1930s indicated that a similar camel trip would cost around C$0.217/ton-km. (see Ho Yang-ning, Cha-sui-meng-min ching-chi ti chieh-p'ou [An analysis of the economy of the Charhar-Suiyuan- Mongol region][Shanghai: Commercial Press, 1935], p.81).

^{g}This is the average unit cost of cart transport in Northwest China for the 1920s. It was taken from Wang Chien-hsin, tr., Chung-kuo nung-shu (Agricultural handbook of China) 2 vols. (Shanghai: Commercial Press, 1934), 1:193.

^{h}These are average cart rates in Shantung in the 1920s; see Wang Chien-hsin, p. 193.

^{i}This was taken from Buck, LUSV, p. 347 for long-distance junk transport on the Yellow River.

^{j}These are second class freight rates.

^{k}The lower figure of C$0.0124/ton-km. pertains to long-distance junk transport at Tai Hsien (Chiangsu); the higher figure of C$0.0497/ton-km. was the average rate for long-distance junk shipment at Wühsi (Chiangsu)(see Buck, LUSV, p. 347). The weighted average rate of long-distance transport costs at Wühsi, Tai Hsien, and Ch'angshu (all Chiangsu localities) was C$0.029/ton-km.

^{l}The lower figure of C$0.0373/ton-km., taken from Buck (LUSV, p. 347), was the long-distance steamer rate recorded at Ch'angshu (Chiangsu). The higher figure was the average tea shipment cost by steamers on the lower Yangtze in the late 1920s (see Shanghai te-pieh-shih she-hui-chu, Section II, p. 43).

^{m}These are second class rates.

^{n}The lower rate of C$0.0186/ton-km. was the average rice shipment rate between Shenshih Hsien (Chiangsu) and Tungp'u (Chechiang)(Chang P'ei-kang and Chang Chih-i, Chechiang-sheng shih-liang chih yün-hsiao [Transportation and marketing of food (grains) in Chechiang][Ch'angsha: Commercial Press, 1939], p. 109). The higher figure of C$0.0336 was for rice transport between Tsungchiang and Shanghai (Institute of Social and Economic Research, Tsungchiang mi-shih t'iao-ch'a [Survey of the Tsungchiang rice market][Shanghai: Institute of Social and Economic Research, 1936], p. 31).

TABLE 4

RAIL TRANSPORT DEVELOPMENT IN CHINA: SELECTED INDEXES*
(1916-26 AVERAGE = 100)

Year	Passenger[a] Traffic (A)	Total[b] Freight Traffic (B)	Agri. & [c] Livestock Traffic (C)	Carrying[d] Capacity of Freight Train (D)
1905-10[e]	31.10	...	...	...
1911-15[f]	53.61	67.99	...	82.45
1916-20	100.00	100.00	100.00	100.00
1921-25	141.37	130.76	118.10	161.66
1926-30[g]	110.63	72.00	...	109.09
1931-35	165.87	153.59	89.65	170.30

SOURCE: Simplified and modified from Yen Chung-p'ing, <u>Chung-kuo chin-tai ching-chi shih t'ung-chi</u> (Statistics on modern Chinese economic history)(Peking: Scientific Press, 1955), pp. 196, 207-8, 211.

*Excluding Manchuria but including the Manchurian section of the Peking-Mukden railway.

[a] Original units in 10,000 passenger-km.

[b] Original units in 10,000 metric ton-km.

[c] Original units in 10,000 metric tons.

[d] Original units in metric tons.

[e] 1907, 1908, and 1909 average.

[f] 1912 and 1915 average.

[g] 1926-1929 average.

sharp plummeting of the selected indexes. The Peking-Hankow railway alone, for example, suffered damages totaling above C$84 million from 1926 to 1930.[1] Apart from losses in terms of property damage, traffic suspension, and commandeering of rolling stock by militarists, the railway system apparently also suffered from regional trade embargoes frequently imposed by warlords, the increased exaction of transit duties for military finance, and the augmented risk of long-distance shipments during time of war. Recovery of efficiency in the operation of most railway lines showed promising signs in the early 1930s.

Agricultural and livestock traffic on the railways, however, never was able to recover fully. This was because the terms of trade

---

[1] Wang Ch'in-yü, pp. 32-33.

for agricultural products (the ratio of farm export prices to rural import prices) registered steep declines during the depression years of early 1930s, while at the same time stepped-up anti-Communist military campaigns in the countryside and the looming menace of Japanese invasion exercised adverse influences on domestic trade expansion. The much publicized reduction in railway freight rates and the official abolition of the likin (transit duties) since 1931 obviously helped only marginally in stimulating agricultural trade.[1] This suggests why the role of railways in the agricultural development of prewar China must be interpreted within the context of the sequence of highly destablizing political and economic events that engulfed the country in those prewar decades.

---

[2] It appeared that for the most part, the abolition of the likin was only nominal; see the discussion in chapter VI.

CHAPTER III

RAILWAYS AND THE CHANGING PATTERNS OF
DOMESTIC AGRICULTURAL TRADE

Historically, the impetus afforded agriculture by railways is one of the more palpable and immediate benefits that rail transport development conferred on traditional economies.[1] Geographical specialization of agriculture and consequent productivity increases followed quickly, as railways improved the conveyance of low-valued and bulky agricultural products over long distances.[2] In backward economies where markets were segregated by the high cost of traditional transport, railways often provided a strong stimulus to the expansion of interregional trade. Even in regions where water transport was well developed, railways would still be a potent force in transforming the pattern of long-distance trade because of such embodied benefits as speed, dependability, and regularity.

This chapter will be devoted to an examination of how railways affected the growth and the spatial pattern of interregional trade in agricultural commodities during the first three decades of this century in China proper. The first section will briefly describe the growth of interregional trade, and then in the following section the role of railways in the growth process will be highlighted by contrasting the trade patterns before and after the introduction of rail transport. A final section will examine the synchronized development of railways and export trade at major treaty ports. This analysis will help dispel the suspicion that the major effect of railways on trade was diversionary. The conclusion that railways had indeed stimulated domestic agricultural trade in China proper since the beginning of this century falls closely in line with the findings of the next two chapters in this study.

---

[1] Agriculture, rather than industry, appeared to be the prime beneficiary from the expansion of markets associated with the development of railways in many historical cases. The reason seems to lie, first, in the fact that agricultural products are generally of lower value relative to weight and hence of lower cost-bearing capacity in transportation than manufactured products. Second, in the early phases of industrialization, the domestic market for manufactures is largely dependent on the level of farm incomes which in turn are strongly affected by transport development. See, for example, Fishlow, pp. 205-36; Hurd, "Railways and Market Expansion in India;" idem, "Economic Impacts of Railways;" Metzer; and White.

[2] Colin Clark and Margaret Haswell, The Economics of Subsistence Agriculture (New York: Macmillan, 1967), p. 199.

The Growth of Interregional Trade

Although the magnitude of growth in the non-local trade of agricultural products does not always epitomize the degree of structural change in the process of modern economic growth,[1] such a relationship apparently was characteristic of the growth process in early modern China. In the first three decades of the present century, the real value of China's farm output that entered into extraregional trade of one hundred miles or more in distance at least doubled.[2] The exclusion of Manchuria did not substantially change this observed magnitude of expansion in agricultural trade. Table 5 gives an estimation of the long-distance farm trade in Manchuria in the 1930s with the assumption that a substantial proportion of this trade was ultimately exported. The estimated total was C$297.8 million, which leaves total extraregional trade in China proper at about C$1,736 million in 1933. This is about double the conservative estimate for China proper for the early 1900s (C$823 million).

The expansion of interregional trade in prewar China received considerable impetus from the steady rise in the export of farm products. China's farm exports more than doubled from C$228 million to C$479 million (in 1933 prices) over the course of the two decades between 1900-09 and 1920-28; but the increase in farm exports accounted for no more than a quarter of the estimated total increase in long-distance trade. Apparently, the strongest stimulus to the

---

[1] As Simon Kuznets phrases it, modern economic growth implies major structural changes or, specifically, "shift away from agriculture, associated with rise in income per capita during the course of economic growth which led to the identification of the latter with 'industrialization,' 'urbanization,' and associated processes." (Simon Kuznets, Six Lectures on Economic Growth [New York: The Free Press, 1959], p. 58). The increased exportation of agricultural products to distant markets is a natural result of the rise in the size and real income of the non-farm sector and the growth of industries which depend on farm output as raw materials. Historically, however, territories that were under colonial rule did exhibit large shipment of products out of the agricultural, or more often the implanted plantation sector, though structural change of any appreciable magnitude was scarcely in sight. This is also true of autonomous, mainly export-oriented agrarian economies during early phases of economic development.

[2] According to Perkins, extraregional trade of farm products amounted to no more than 400-500 million taels, or 7-8% of total farm output in 1900-1910. In constant 1933 prices this would be about C$823-C$1,029 million. By 1933, however, domestic long-distance agricultural trade was estimated at C$2,034 million, or over 13% of total farm output (see Perkins, Agricultural Development, pp. 119, 136, 289). In other words China's long-distance trade in farm products had grown by 98-146%, or most likely over 100%, between 1900-1910 and the early 1930s. This magnitude of growth should of course be adjusted upward should one compare the change between the 1890s and the 1930s.

TABLE 5

ESTIMATION OF LONG-DISTANCE AGRICULTURAL TRADE IN MANCHURIA (1931-37 AVERAGE)

Item	Output ('000 metric tons)	Percent Exported	Export ('000 metric tons)	Unit Price (C$/metric ton)	Total Value (million C$)
Soybean	4,302.3	73.6*	3,165.0	78	246.8
Kaoliang	4,082.8	2.3	106.1	56	5.9
Millet	2,894.4	10.9	324.3	72	23.4
Corn	1,964.4	16.5	313.3	58	18.2
Wheat	1,026.0	2.5**	38.9	90	3.5
Total					297.8

SOURCES: Output figures from Kungtu C. Sun, The Economic Development of Manchuria in the First Half of the Twentieth Century (Cambridge, Mass.: Harvard University Press, 1973), p. 58; percent export figures, unless otherwise specified, were taken from Li Wen-chih and Chang You-i, eds., Chung-kuo chin-tai nung-yeh shih tz'u-liao (hereafter CKNY)(Historical materials on agriculture in modern China) 3 vols. (Peking: San-lien Book Co., 1957), 2:231; price data were from Perkins, Agricultural development, p. 288.

*Derived from Kungtu Sun, p. 59, as total of beans exported and beans used in the production of beancakes and bean oil. Output to raw material ratios are given in ibid., p. 16.

**Total of wheat and processed wheat products.

growth in domestic trade emanated not from changes in foreign demand, but from growth in the domestic demand for farm products. Developing from a very small base in the 1890s, output of China's agricultural processing industries amounted to an estimated C$12 billion by 1933, with net value added contributing to about 4% of the country's Net National Product.[1] Side by side with this industrialization process was the growth in urban population, which between the 1900s and the mid-1930s had added more than 10 million people to the country's large cities (cities with more than 100,000 people by 1958).[2] As a result, the large city population in China increased more than 70% in the first three decades of the present century.

---

[1] Calculated from data in Liu Ta-chung and Yeh Kung-chia, The Economy of the Chinese Mainland (Princeton: Princeton University Press, 1965), pp. 66 (table 8), 146 (table 39), 155 (table 45). The main categories included in the computation were textile products (woolen, cotton, and silk), edible vegetable oils, cigarettes, wheat flour, and milled rice.

[2] Perkins, Agricultural Development, p. 128.

Both the development of agricultural processing industries and the sizable increase in urban population imply that the farm sector was burdened with the need to generate an ever larger agricultural surplus. Moreover, the demand for food by urban dwellers received additional reinforcement from a probably steady rise in urban real income.[1] Given the traditional technology of production, the capacity of China's peasant economy to meet these marked rises in demand would indeed be heavily strained. Perkins, in fact, suggested that much of the increased urban population was fed not by concomitant rises in the domestic supply of foodgrains, but by the increases in foreign imports.[2] However, he may have erred by overlooking the fact that a substantial portion of the imported grains was consumed in the food-deficit regions along the southern and southeastern seaboard. The fast growing industrial and population centers of North and Central China, including Shanghai, were more likely dependent on domestic than on foreign sources of supply, except perhaps during years of abnormal home shortages.[3] As late as the early 1930s, imported grain probably accounted for no more than 0.2% of the

---

[1] There is no published study which attempted an estimation of changes in urban real income during this time period. A rough conjecture may start with the estimated Gross Domestic Product (GDP) accounted for by the non-farm sector in the years 1914-18 and 1933, the only years for which China's GDP has been estimated by industrial origin. According to Perkins, these amounted to C$18.5 billion and C$26.5 billion (in constant 1957 prices) for 1914-18 and 1933 respectively (Dwight Perkins, "Growth and Changing Structure of China's Twentieth-Century Economy," in China's Modern Economy, ed. Dwight Perkins, p. 117). After adjusting for depreciation and indirect taxes, urban per capita income may be thus derived and can be shown to have increased by more than 6% during the 15-20 year interval.

Some scattered information on the incomes of certain professional groups in large cities also helps shed light on this issue. The daily wages of unskilled men working with Peking's carpenter and mason's guild were about 200 copper cashes in the 1890s (quoted S. Gamble in Wagel, p. 270) and about 65-90 silver cents in the late 1920s (quoted S. Gamble in J. B. Condliffe, China To-Day: Economic [Boston: World Peace Foundation, 1932], p. 107). Adjusted for rises in prices and changes in the copper-silver exchange rate, these figures reveal over 20% increase in the real wages of these urban workers in the thirty years since the 1890s. If it can be assumed that this scale of increase was representative of unskilled workers, average incomes of professionals and skilled workers would probably have registered even larger increases.

[2] Perkins, Agricultural Development, p. 155. Perkins' speculation was based on the matched increases in urban consumption demand and the import of foodgrains, mainly rice and wheat.

[3] According to Murphey, Shanghai in the late 1920s and early 1930s actually drew most of its rice supply from a local region of no more than 80 miles in radius (Rhoads Murphey, Shanghai, Key to Modern China [Cambridge, Mass.: Harvard University Press, 1953], pp. 139-51). Though this need not mean that an umland of 80 miles in radius was sufficient to support a metropolis of 3 million since the umland itself was dotted with large regional collection centers of rice, this is sufficient to de-emphasize the role of imported rice in the case of Shanghai. Shanghai, however, was a major importer of foreign wheat. But most of this wheat was raw material to the flour-milling industry whose product was distributed to other ports.

country's total consumption demand for cereals.[1] As Myers has suggested, Chinese agriculture was able to keep pace with the rapid growth in rural and urban population during the four decades since the 1890s without excessively relying on imported food.[2] As a result, China was able to shift her import composition gradually in favor of producer's goods as the share of food in total import declined.[3]

A basic factor which enabled China's agrarian economy to cope successfully with rising demands in the prewar era was, as Myers argued, the spread of agricultural commercialization and the improvement in transport conditions.[4] Agricultural commercialization, in fact, was intimately entwined with developments in the transport sector. As will be seen in chapter V, railway construction provided especially a strong drive for the expansion of commercial cropping in prewar China. It is thus not surprising to find that within China proper, the fastest pace of urban and perhaps industrial growth between the 1890s and the 1930s occurred in North China, where railways signified an unparalleled advance in long-distance transport technology in the absence of an extensive network of navigable waterways.[5] As a rough indicator of urban-industrial growth, total exports from North China multiplied nearly thirty-eight-fold between 1900 and 1936, from about 3 million H. K. taels to over 120 million H. K. taels.[6] In comparison, total exports from the Yangtze provinces (Central China) increased a moderate 180%, and that from the rest of South China a mere 30% during the same time period.[7] Urban population in the North China provinces of Shantung, Hopei, Honan, Shanshi, and northern Anhui nearly doubled between 1900-10 and 1938, as compared

---

[1]Cheng Yu-kwei, Foreign Trade and Industrial Development of China (Washington, D.C.: The University of Washington Press, 1956), p. 257, note 62.

[2]Myers, "The Commercialization of Agriculture," p. 174.

[3]Cheng Yu-kwei, pp. 33-34.

[4]Myers, "Commercialization of Agriculture."

[5]North China accounted for only about 15% of the total mileage of navigable waterways in China, and about 87% of these were suitable for junk traffic only (see de Fellner, pp. 192-94). Also, as mentioned before, about 35% of the time in a normal year was considered unsuitable for navigation in North China's rivers.

[6]Calculated from data in C. Yang et al., Statistics of China's Foreign Trade During the Last Sixty-Five Years (Shanghai: National Research Institute of Social Sciences, Academia Sinica, 1931); and Cheng Yu-kwei, p. 49, table 18. One (Haikwan) tael was equivalent to about C$1.5 in the 1930s.

[7]See n. 6 for sources of data in the calculation.

with about 45% increase for both the Yangtze provinces and others in South China, including Yunnan and Hueichou.[1]

Railway development in North China and in Manchuria appeared to have afforded a unique opportunity for the expansion of North China's urban-industrial sector which before was very much constrained both by the long-distance (and hence high transfer cost) separating it from the food-surplus central Yangtze provinces and the lack of domestic sources of raw material supply for the agricultural processing industries. The railways in North China stimulated the production of industrial crops at the same time as they reduced the cost of long-distance transportation.[2] Equally important, however, was the contraction of space made possible by the construction of railways. Whitney estimated that within the territory served by railways, time distance in China during the first thirty years of this century contracted by 97%.[3] Railways thus became a potent force in the internal integration of the Chinese economy, and the observed expansion in interregional trade closely reflected this fact.

The Spatial Configuration of Domestic Trade Expansion

Before the advent of mechanical transport, the high cost of land transportation virtually prohibited the commercial flow of low-valued, bulky farm products overland for any considerable distance. Trade along land routes was confined to a few luxurious commodities, such as silk and tea, which accounted for only a very small part of aggregate domestic commerce. The dominant flow of trade followed the pattern of natural and artificial waterways with which South China (including Central China) was richly endowed. The Yangtze and its tributaries in particular constituted the main artery of China's internal trade for centuries.[4] The building of

---

[1] Based on Perkins, <u>Agricultural Development</u>, pp. 292-95.

[2] For an estimation of railway-induced decline in transfer cost for a sample of rural communities see chapter IV below. Appendix E contains a general estimation of the average decline in long-distance transfer cost in prewar China since the 1900s.

[3] Joseph Whitney, <u>China: Area, Administration, and Nation Building</u>, Research Papers, no. 123 (Chicago: University of Chicago, Department of Geography, 1970), pp. 45-47.

[4] The basic pattern of grain trade, for example, probably had changed little from the 12th century; see Ch'uan Han-seng, "Production and Distribution of Rice in Southern Sung," in <u>Chinese Social History</u>, ed. E-tu Zen Sun (Washington, D.C.: American Council of Learned Studies, 1956), pp. 222-33.

railways and the momentum it imparted to industrial and commercial development in such Yangtze ports as Shanghai, Nanking, Hankow, and Ch'angsha actually reinforced the importance of the Yangtze in domestic trade.

The prominent position of the Yangtze in domestic trade stemmed mainly from the fact that it serves a territory which was the major rice-exporting region of traditional China. The chronically food-deficit areas of the country, on the other hand, spread mainly along the eastern and southeastern seaboards, which were several hundreds of miles or more by water from the exporting regions (see figure 4a). The flow of rice from the surplus central Yangtze provinces, and even from Ssuch'uan at times of poor harvest in the lower Yangtze basins, amounted to as much as 8-13 million shih (0.6-1 million tons) per annum in the early 1700s.[1] However, the volume of this traffic had probably diminished somewhat since then.[2] At least part of this rice was transferred north via the Grand Canal to meet the needs of Peking and Tientsin, the only truly large cities in the north before the turn of the century.[3] Peking, the Ch'ing imperial capital, absorbed an annual inflow of 3-4 million shih (230,000-310,000 tons) of tribute rice conveyed through this channel.[4] Thus the Grand Canal was the major highway of commerce between North and South before the steamer gradually replaced it with a maritime route along the coast in the latter half of the nineteenth century.

Besides the Yangtze course, rice was also carried by the Han River upstream to the deficit region of Hsian and by the Hsi River downstream to Kuangtung, the latter amounting at times to two million shih (150,000 tons) or more.[5] The agricultural trade along these routes, particularly that carried through Hsi River, however, never

---

[1] Ch'uan Han-seng and Richard Kraus, Mid-Ch'ing Rice Markets and Trade (Cambridge, Mass.: Harvard University Press, 1975), p. 77.

[2] This possibility arises from a much smaller estimate by Perkins for the 1930s; see ibid., pp. 77-78.

[3] For data consistency, the revised urban population data in Perkins (Agricultural Development, pp. 292-95) were used.

[4] Ch'uan and Kraus, p. 64. The amount of tribute rice carried north annually probably also declined between the 1700s and the 1900s, for at least one record put the amount at only 1.6 million piculs (about one million shih) in the early 1900s; see Morse, p. 326.

[5] Ch'uan and Kraus, p. 77. The Hsi River route, it may be noted, was the only communication and trade channel between the Yunnan and Kueichou region and the Pearl River delta region prior to the construction of the French Indochina railway which opened the Yun-Küei highlands to external trade through Haiphong in Vietnam. The Han River route was an old trade and postal route which supplied the need of the old dynastic capital of Hsian.

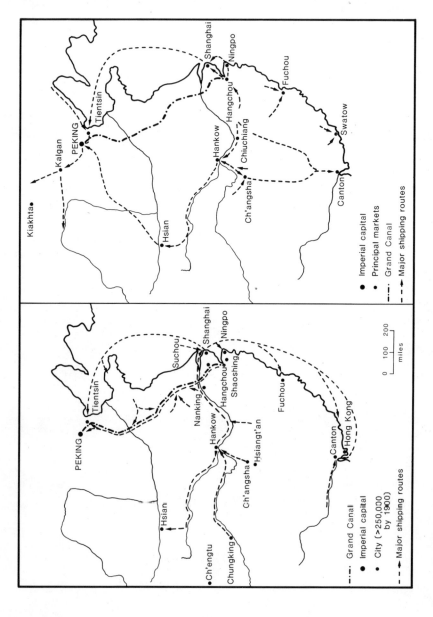

Fig. 4. Long-distance agricultural trade in China, c. 1870: (a) rice trade (after Perkins, map VII.3); (b) other crops, mainly tea (from information in Morse, pp.313-26 and Torgasheff, pp.61-62)

exhibited much diversification apart from grains. The Yangtze, in contrast, commanded not only the principal channel of grain shipment but also that of most other commercial crops which ever played a visible role in the domestic commerce of pre-modern China. One major factor behind this was the richness in such trade staples as silk, tea, and cotton in the Yangtze basins, and the highly regionalized cultivation of these crops only contributed to an interregional commerce fairly internalized to the greater Yangtze system. The coming of the steamship and the opening of the Yangtze to foreign shipping after the 1860s gave considerable impetus to the production of these crops for export and greatly strengthened the river's domination in agricultural trade.

Among the commercial crops traded on the Yangtze, tea, which was China's biggest export item in value terms up to the 1890s,[1] was the most important. Prior to the steamship era, considerable amounts of the tea produced in the hilly countrysides of Huküang (Hupei and Hunan), Anhuei, and Kiangsi were conveyed to Canton for sale to British merchants via two routes: one upstream on the Hsiang River and over the Cheling Pass; the other upstream on the Kan River and over the Meiling Pass; both routes then followed the Pei River downstream to Canton (figure 4b).[2] Tea from the middle Yangtze provinces was also carried north, mainly by Shanhsi merchants after the opening of the Kiakhta market in Siberia in 1727, first via the Han River to Hsian and then overland to Tientsin, Kalgan, and across Mongolia to Kiakhta (the Kiakhta route)(figure 4b).[3] However, tea trade on these routes of shipment was quickly displaced after the 1840s by the fast growing tea market at Hankow, where first the British and then the Russians (beginning in the 1880s) directly exported the tea by steamers. By the 1870s, tea directly exported from Hankow already accounted for some 37% of the total tea export of China;[4] the latter having grown as much as five-fold (in quantity) in the thirty years since the 1840s.[5] The growth in tea trade

---

[1] Cheng Yu-kwei, p. 15.

[2] Morse, pp. 313-15.

[3] Liu Ts'ui-jung, "Trade on the Han River and Its Impact on Economic Development, ca. 1800-1911," (Ph.D. dissertation, Harvard University, 1974), p. 47; see also Morse, pp. 323-25.

[4] Based on data in Maritime Customs, Reports and Returns of Trade for 1878 (Shanghai, 1879), Part I, section on Hankow; and C. Yang et al., p. 35.

[5] Total tea export to England, the sole foreign purchaser of Chinese tea before the 1840s, amounted to 350,000 piculs in 1848-50. Total tea export had grown to about 1.8 million piculs by 1876, or more than five times the level in

figured significantly in the expansion of interregional trade because of the much higher value of tea compared to rice.[1] Hankow also became an increasingly important market for the exports of cotton and sesame, drawing supplies mainly from the Han River area. These crops, however, did not become important in domestic trade before the railways greatly expanded their national markets.

Beginning in the early 1900s, the evolution of a national network of railways brought in much diversity, both in the orientation and composition of trade, to the once simple and river-dominated pattern of China's domestic agricultural commerce. First, the completion of the Peking-Hankow railway introduced a clear north-south dimension to the old trade pattern characterized by the country's large west-east flowing rivers (figure 2). Then the Tientsin-P'uk'ou (opposite Nanking) railway reinforced this north-south interaction by taking over the role once played by the Grand Canal and by engaging in beneficial competition with coastal shipping by steamers. These two trunk lines constituted the backbone of a railway network which, though not imposing in size and density, was able to evoke significant shifts in the frontier of accessibility under water transportation. Following in the wake of this transport development was the regionalized cultivation of a variety of industrial crops which soon formed the core of the expanding interregional commerce. We start by looking at the spatial impress railways have left on the flow pattern of interregional trade; closer exmination of empirical evidence will be pursued in the following two chapters.

Figures 5a through 5h provide a visual impression of the more conspicuous flow patterns of major commercial crops in China proper in the early and mid-1930s. Scarcity of detailed regional trade statistics precludes the delineation of the trade patterns of marketed farm products beyond the identification of certain principal routes of shipment. The marked concentration of food-processing and cotton textile industries in Shanghai, Tientsin, Hankow, and Ch'ingtao, however, greatly simplified the domestic trade pattern of

---

1848, (See Boris P. Torgasheff, <u>China as a Tea Producer</u> [Shanghai: Commercial Press, 1926], pp. 152, 165).

[1] Silk, though surpassing tea to become China's major export (in value terms) by the 1890s, was not conspicuous in long-distance trade mainly because of the propinquity of the sites of production and market of exportation. Most of China's silkworm cultivation concentrated in the Yangtze and Pearl River delta areas where the two major export markets, Shanghai and Canton, were located. The silk from Shantung was also produced in coastal regions close to export markets, though perhaps part of the Shantung silk was shipped to Canton for export along a coastal route.

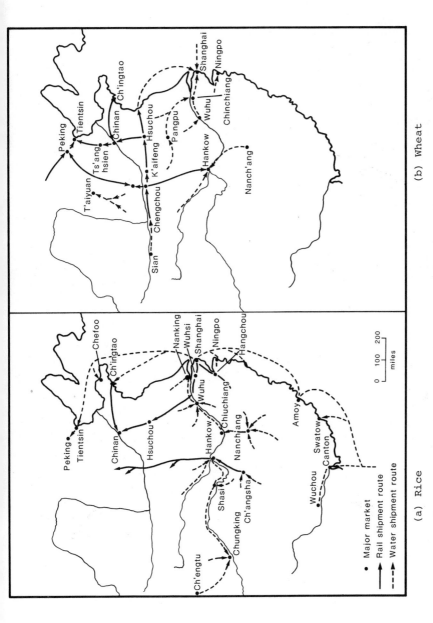

(a) Rice (b) Wheat

Fig. 5. Long-distance agricultural trade in China, c. 1935

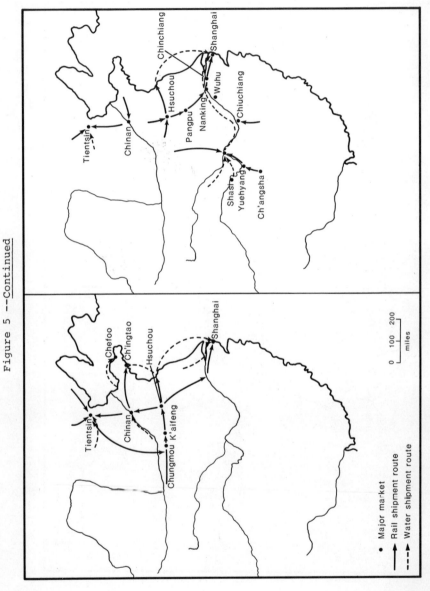

Figure 5 --Continued

(c) Peanut

(d) Sesame

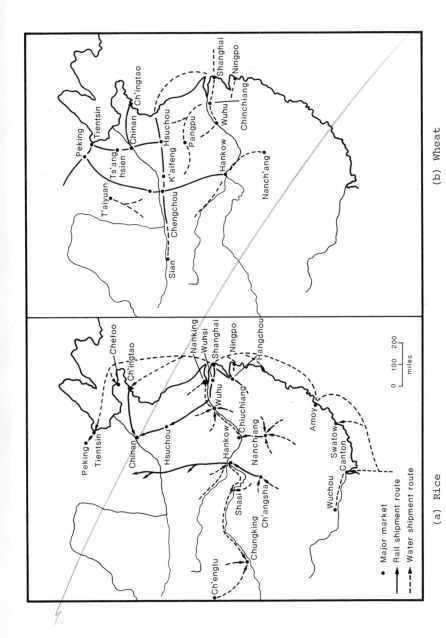

Fig. 5. Long-distance agricultural trade in China, c. 1935.

(a) Rice   (b) Wheat

Figure 5 --Continued

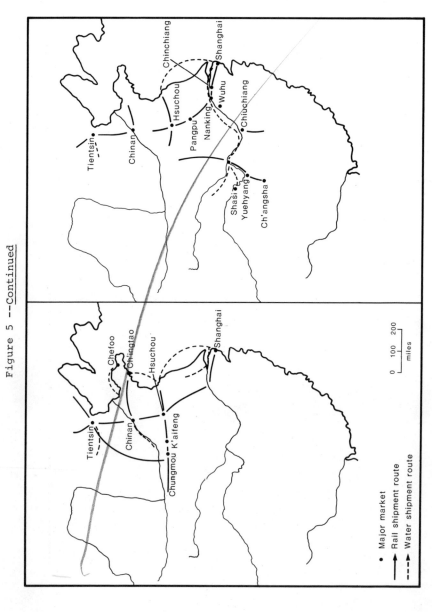

(c) Peanut

(d) Sesame

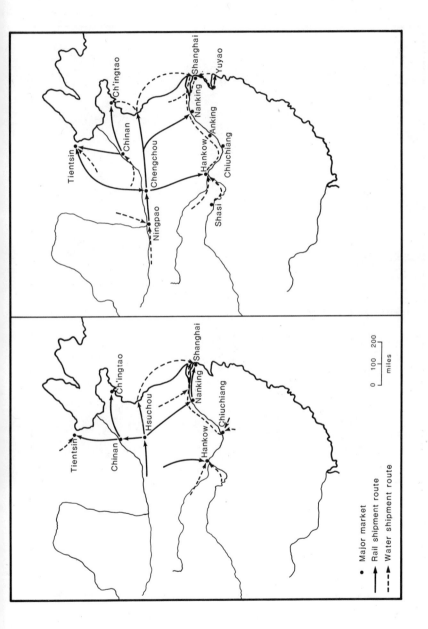

(e) Soybean   (f) Cotton

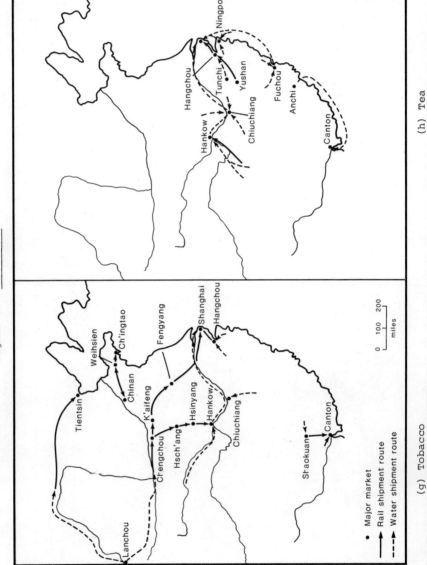

Figure 5 --Continued

(g) Tobacco

(h) Tea

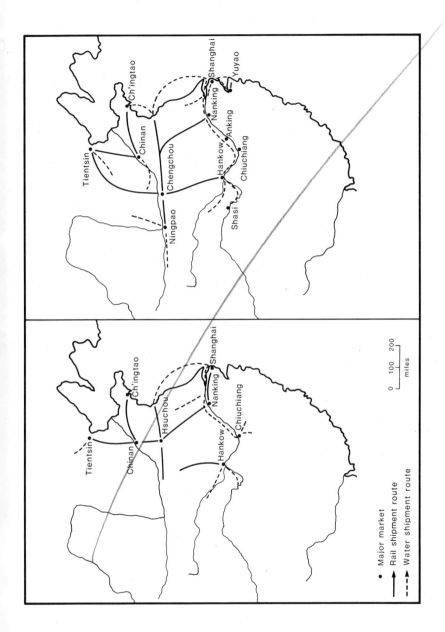

(e) Soybean

(f) Cotton

Figure 5 --Continued

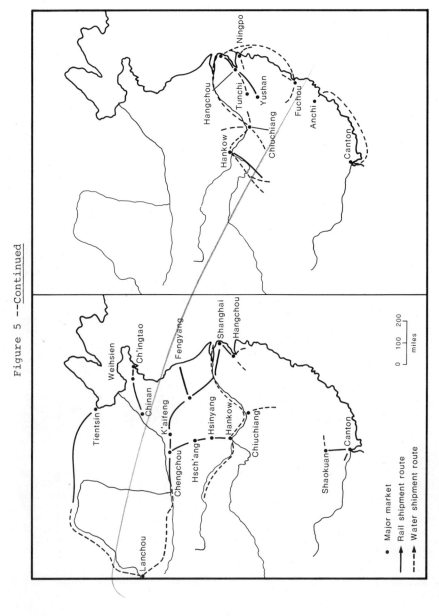

(g) Tobacco

(h) Tea

Figure 5 --Continued

SOURCES: Figures 5a through 5h are impressionistic rendering of data in a number of sources. The more important ones are listed below.

Fig. 5a: Shen Tsung-han, Chung-kuo nung-yeh t'zu-yuan (Agricultural resources of China) 3 vols. (Taipei: Chung-hua wen-hua chu-pan-shih-yeh wei-yuan-hui, 1953), 1:146 and map 16; Perkins, Agricultural Development, p. 152, map VII.4; Economic Research Unit, Bank of China, Mi (Rice [in China])(n.p., 1937), pp. 106-217 passim.; Chang P'ei-kang and Chang Chih-i, pp. 13-36, 87-117 passim.; Chang P'ei-kang, Chianghsi shih-liang wen-t'i (The Chianghsi grain problem) (Shanghai: Institute of Social and Economic Research, 1935), pp. 11-12, 39-40, 44-49; Chang Chih-i and Ou Pao-san, Fuchien-sheng shih-liang chih yün-hsiao (Grain marketing and transportation in Fushien Province)(Shanghai: Commercial Press, 1938), pp. 5-26; Chang P'ei-kang, Kuanghsi liang-shih wen-t'i (The Problem of grains in Kuanghsi Province)(Shanghai: Commercial Press, 1938), pp. 53-74; Research Institute, Chiao-t'ung University, Yuan-chung tao-mi chih t'iao-ch'a (A survey of rice [Production and Marketing] in Central Anhuei Province)(n.p., 1936), chapter 2.

Fig. 5b: Shen, 1:146-48, map 17, 2:86-88, and maps 24-25.

Fig. 5c: Foreign Trade Bureau, Ministry of Industry, Hua-sheng (Peanuts [in China])(Shanghai: Commercial Press, 1940), pp. 42-52, 120-48 passim.; Shen, 2:190-193, maps 56-57; and "Production and Export of Groundnuts in China," Chinese Economic Journal 19 (September 1936):257-68, (October 1936):374-95.

Fig. 5d: Foreign Trade Bureau, Ministry of Industry, Chih-ma (Sesame [in China])(Shanghai: Commercial Press, 1940), pp. 7-28, 32-41, 103-104 passim.; and Shen, 2:193-94, maps 58-59.

Fig. 5e: Shen, 2:169-74, maps 52-53; Chou Chih-hua, Chung-kuo chung-yao shang-pin (Important Commercial Articles in China)(Shanghai: Hua-t'ung Book Co., 1930), pp. 126-34; Tsao Lien-en, "The Marketing of Soya Bean and Bean Oil (in China)," Chinese Economic Journal 7 (September 1930):941-71.

Fig. 5f: Fong Hsien-t'ing (H. D. Fong), Chung-kuo chih mien-fang-chih yeh (Cotton textiles industry in China)(Shanghai: Commercial Press, 1934), pp. 21-84 passim.; Department of Agricultural Economics, Nanking University, Yu, Ngo, Yuan Kan shih-sheng chih mien-ch'an yün-hsiao (Cotton marketing in the four provinces of Honan, Hupei, Anhui, and Chianghsi)(Nanking: Nanking University Press, 1936), pp. 37-80 passim.; Shen, 3:47-53, maps 62-63; "Cotton Production in Hupei," Chinese Economic Journal 13 (September 1933): 356-63; "Raw Cotton Trade in Shanghai," Chinese Economic Journal 3 (July 1928):681-92.

Fig. 5g: Foreign Trade Bureau, Ministry of Industry, Yen-yeh (Tobacco [in China])(Shanghai: Commercial Press, 1940), pp. 9-36, 100-105, 109-28; Chen Han-seng, Industrial Capital and Chinese Peasants (Shanghai: Kelley & Walsh, 1939), pp. 16-32 passim.; Shen, 3:10-15, maps 60-61; "Tobacco Production and Marketing in China," (Part I) Chinese Economic Journal 16 (March 1935):407-20, (Part II) (May 1935):531-43; and Nan Pin-fong, Honan ch'an-yü ch'ü chih t'iao-ch'a pao-kao (A survey report of tobacco producing areas in Honan)(Nanking: Nanking University Press, 1934), pp. 1-2, 6-10.

Fig. 5H: Chu Mei-yu, Chung-kuo cha-yeh (Tea in China)(Shanghai: Chung-hua Book Co., 1937), pp. 46-138 passim.; Chu T.-H., Tea Trade in Central China (Shanghai: Kelley & Walsh, 1936), pp. 6-256 passim.; "Tea Production and Trade in Chechiang," Chinese Economic Journal 14 (May 1934):521-36.

most industrial crops. For most of these crops Shanghai secured a primacy in the hierarchy of trade ports as it commanded the Yangtze trade hinterland. Tientsin, Hankow, and Ch'ingtao formed a second tier in the hierarchy.[1] They were terminal markets of primary trade flows originated in or transhipped from smaller regional markets, the more important of which included Paot'ou, T'aiyüan, Chinan, Chengchou, Hsüchou, Nanking, Hangchou, Ningpo, Chiuchiang, and Ch'angsha.

The mid-1930 pattern of grain trade, dominated by rice and wheat, was much more complex than that of other commercial crops (figures 5a and 5b). This mainly resulted from the much higher degree of dispersion of production and consumption points in the case of foodgrains. Grain trade also offered the most discernible departure from the pattern that prevailed prior to the 1880s. Instead of being transported over long distances to deficit coastal areas, rice in the upper and middle Yangtze provinces of Ssuch'uan, Hupei, and Hunan had largely become an intraregionally traded commodity by the mid-1930s.[2] This change made the teeming population in the coastal lowlands of Kuangtung and Fuchien increasingly dependent on imported rice and other foodstuffs (e.g., flour) from overseas, though Shanghai was perhaps immune to this pressure by acquiring sufficient supply from nearby provinces.

The shipment of rice, as before, followed a largely water course. However, even for this low-priced bulky commodity, railways acquired some importance as a carrier. Rice supplied to deficit communities along the Peking-Hankow railway was handled exclusively by rail. The Tientsin-P'uk'ou line, while engaging in tripartite

---

[1] This hierarchical structure conformed closely to the ranking of these cities in their shares of the national totals of foreign and domestic trade. Shanghai alone accounted for some 55% of total foreign trade of the country and 38% of the domestic trade averaged for the 1925-35 period, while Tientsin, Hankow, and Tsingtao together took up 19% of the foreign and 20% of the domestic trade (see Murphey, p. 121, table IV). Canton, though sharing 6% of foreign and 4% of domestic trade, played an almost insignificant role in the exportation of the industrial crops here discussed (except for silk which, as mentioned, was not a conspicuous item in long-distance trade).

[2] Population growth and urbanization in these regions were likely responsible for this change (see Perkins, Agricultural Development, pp. 153-55), but the conversion of paddy fields to other lucrative crops might also be an important factor. At least one writer mentioned this as a major cause of Hankow's reversion from a rice-exporting port to a rice-importing market (see Chu Chien-pang, p. 85). Data on paddy acreage also corroborated the decline of rice cultivation in Hunan, probably by more than 30% between 1914-18 and 1931-37, while at the same time the acreage occupied by such industrial crops as peanut, cotton, and sesame greatly expanded (see Perkins, Agricultural Development, pp. 249, 259, 261, 264). Though data on Hupei are lacking, it is very likely that a similar trend occurred in Hupei as in Hunan, particularly in more accessible areas near Hankow.

competition with the Grand Canal and coastal steamers, also functioned viably as an important rice-carrier to urban centers along the railway track. Even in regions where direct water competition was severe, railways appear to have fared unexpectedly well as a transporter of rice. For example, rice collected at Ningpo (Chechiang province) to be redistributed to adjacent deficit prefectures was usually carried by the Hu-Hang-Yung (Hangchou-Ningpo) railway.[1] The Shanghai-Hangchou line, though traversing a territory crisscrossed by numerous natural and artificial waterways, was reportedly active in the carriage of surplus rice from northern Chechiang to Shanghai and of Shanghai rice to Hangchou.[2] Upon the completion of its Chechiang section in 1934, the Chechiang-Chianghsi railway soon evoked the shift of a large portion of rice from a traditional river route to railbound traffic, precipitating a rivalry which forced river junks to abolish many of their common fraudulent practices in consigning rice.[3] Another notable case was provided by the Nanch'ang-Chiuchiang railway in Chianghsi. A goodly part of the rice conveyed to Chiuchiang from Nanch'ang, the principal provincial rice-collecting center, was handled by rail, even though sending the rice by a parallel water route on the Poyang Lake would have saved freight cost by as much as 30%.[4] Apart from rice, wheat was probably the only grain crop that entered perceptibly into long-distance trade. The emergence of a sizable wheat trade owed much to the rapid advance of the flour-milling industry during World War I, but the flow must have dwindled considerably during the 1920s throughout which period the industry stagnated.[5] The post-1920 recession of the flour industry was probably precipitated by increased foreign competition. However, the persistent problems of uncertain domestic wheat supply due to repeated crop failure, low quality of home-grown wheat, and the high cost of internal transport militated against the revival of domestic wheat trade, even after 1932, when the flour

---

[1] Chang P'ei-kang and Chang Chih-i, p. 11.

[2] Ibid.

[3] Ibid., also p. 100. These practices included larceny and the adding of water to rice to make up the original volume of the cargo. These acts were facilitated by the "open boat" method commonly practiced in rice consignment, i.e., rice was simply lumped in open rice junks. It may be added that railways were generally entrusted with the conveyance of high-value husked rice while rice junks were favored for shipping unhusked grain (ibid., p. 93).

[4] Chang P'ei-kang, Chianghsi shih-liang wen-t'i, pp. 181-82. It should be noted that railway's successful capture of the bulk of the commerce implied that the real cost of rail shipment was lower than that incurred by junks.

[5] John K. Chang, pp. 43-44.

industry received stimulation from tariff protection.[1] A rough estimate would put the size of wheat trade within China Proper in the early 1930s at no more than half a million metric tons per year.[2] Of this amount, probably some 80% were conveyed by railways.[3] Of particular importance in the carriage of wheat were the Peking-Hankow, Lung-Hai, and Tientsin-P'uk'ou railways, which carried the bulk of the wheat produced in the Wei River and Yellow River territories to the flour-milling centers of Tientsin, Chinan, and Hankow.[4] Nevertheless Shanghai, the capital of the flour industry, relied heavily on imported wheat. Only a small amount of the wheat produced in Anhuei and Chiangsu, shipped by rail and junks alike, was drawn into the Shanghai market.[5]

Like wheat, a large portion of the non-grain industrial crops that formed the core of China's domestic agricultural commerce in the 1930s was scarcely in existence before the 1880s. To be sure, the growth of this commerce was in close pace with industrial expansion and rising overseas demand over the entire course of the early 1900 decades, but the locational affinity of commercial cropping to the railway network raises the question whether rail transport development actually set precedent conditions for industrial growth. This seems to be the case for peanuts and tobacco, if only not as explicit in the case of other crops.[6] Whatever the cause, the close spatial alignment between the production pattern of

---

[1] Ibid.; see also Shen, 1:147. The high cost of transporting wheat here mentioned was due not only to the relatively costly shipment by rail but also to the long distance over which the grain had to be carried by traditional means to rail-heads. Thus it was said to be cheaper to send wheat all the way from North America than from Shenhsi to Shanghai (see "Chinese Flour Industry," Chinese Economic Journal [February 1931]:107).

[2] This was derived from the reasonable assumption that most of the wheat engaged in interregional trade was absorbed by the modern milling industry. Liu and Yeh (p. 526) estimated the amount of wheat consumed by the industry at about 34.8 million piculs for 1933, from which the net import of 21.1 million piculs, and the Manchurian trade of 4.7 million piculs must be deducted to arrive at the trade volume within China Proper, 9 million piculs or 450,000 metric tons (see Liu and Yeh, pp. 428, 444).

[3] In 1935 the Chinese National Railways carried a total of 823,000 metric tons of wheat (Yang, p. 120), of which probably half can be considered as original tons (based on inference from the network of rail connections and shipment origins of the grain).

[4] See "Flour Industry in Shantung," Chinese Economic Journal 15 (September 1934):328-37, especially p. 331.

[5] Shen, p. 147.

[6] This is further discussed in chapter V below.

major industrial crops and the railway network in North China accorded railways natural advantage as a carrier for these crops.

The principal highways of commerce, again, were represented by the three trunk lines: the Peking-Hankow, Lung-Hai, and Tientsin-P'uk'ou railways. In addition, the Ch'ingtao-Chinan line in Shantung emerged as an important regional carrier of certain industrial crops, which was mainly a result of the rapid development of the port of Ch'ingtao and the industrial center of Chinan, as well as the special promotion effort launched by tobacco companies at expanded production of the tobacco crop at sites along the railway line.[1] Except for tobacco, the Ch'ingtao-Chinan line served mainly as a feeder which channelled agricultural traffic from the other trunk lines through the port of Ch'ingtao for export or processing.

Of the three main trunk lines that served North China outside the Shantung peninsula, the Lung-Hai line in particular traverses a territory abundant in the production of peanuts, cotton, and soybeans. The line's parallel course with the Yellow River scarcely gave rise to the menace of water competiton. On the contrary, adverse navigating conditions in the middle stretch of Yellow River actually helped consolidate the monopoly enjoyed by the railway in all eastbound freight traffic from the Wei River basin, southwestern Shanhsi, and west-central Honan.[2] Even cargoes from Kansu and Ch'ingtao, featuring such goods as tobacco, hides, wool, and herbs, passed through this railway route to Shanghai and Tientsin for export.[3] Freight commerce on the Lung-Hai was afforded fresh stimulus in 1932, when through freight traffic and sea-rail intermodal transport services were introduced;[4] the latter directly

---

[1] For further discussion of this theme see chapter V.

[2] The Yellow River, despite its great length, was traditionally not an important highway of commerce. From T'ungkuan eastward to Chengchou, where the river runs parallel to the Lung-Hai railway, the Yellow River descends a steep gorge, carrying with it a silt load from the Loess Plateau heavy enough to choke its outlets to the sea and produce elevated river beds conducive to recurrent, calamitous flooding. The navigable season of the river is also limited to seven months, and then by shallow junks or rafts only (see Wang Kwang, Chung-kuo shui-yün chih [River shipping in China][Taipei: Chung-hua ta-tien p'ien-yin-hui, 1966], pp. 362-64; and T. R. Tregear, An Economic Geography of China [New York: Elsevier, 1970], pp. 60-61).

[3] See Wang Ching-wang, p. 43. Another important route of shipment for agricultural and livestock products from northwestern provinces followed a river course down the Yellow River from Lanchou to Paot'ou (by rafts) and then by the Peking-Suiyuan railway to Peking.

[4] Ministry of Railways, T'ieh-lu nien-chien (Railway yearbook)(Nanking, 1935), part II, pp. 909-11.

connected the Lienyunkang terminal of the Lung-Hai railway with
Shanghai and Ch'ingtao through regular steamer services. The
Peking-Hankow railway, on the other hand, was engaged in a
north-south commerce that represented the bulk of the Hankow-bound
sesame and tobacco from Honan and the Tientsin-bound cotton from
northern Honan and western Hopei. But the monarch in cash crop
commerce was the Tientsin-P'uk'ou line, which was well blessed from
the beginning by its locational advantage as a direct communication
link between Shanghai and Tientsin, the nation's two largest
commercial centers. Table 6 displays the relative shares of the
four most important trunk lines in the total delivered 1935 tonnage
of the major commercial crops. It can be seen that the
Tientsin-P'uk'ou railway led the others with 21% of the national
total (China proper only), followed by the Peking-Hankow,
Ch'ingtao-Chinan, and Lung-Hai lines, in that order.

TABLE 6

RAILWAY TONNAGE IN COMMERCIAL CROPS (1935):
RELATIVE SHARES BY MAJOR TRUNK LINES

Crop	Percent of Total Railway Tonnage				Total Four Lines	National Total ('000 m. tons)
	Peking-Hankow	Tientsin-P'uk'ou	Lung-Hai	Ch'ingtao-Chinan		
Peanut	1	17	14	45	77	234
Sesame	34	34	4	0	72	169
Soybean	29	31	4	10	74	444
Cotton	12	16	18	9	55	747
Tobacco	26	10	1	45	82	106
Tea*	13	9	4	2	28	105
Total	18	21	11	15	65	1,805

SOURCE: Yang Hsiang-nien, pp. 120-121.

*The shares of the Hu-Hang-Yung (Hangchou-Ningpo) and Hupei-Hunan railways
were 21% and 16%, respectively.

The importance of the national railway system in the
interregional transportation of commercial crops may be envisaged
from an estimation of the shares of the railways in the total
interregional trade flows. As with the case for wheat, total
volume of interregional trade for an industrial crop is estimated
as total consumption by the modern factory sector, plus the volume
of net export (export minus import). Based on the patterns of

production and consumption summarized in figure 5, reasonable estimations of the original tonnages of specific industrial crops carried by the railway system can also be derived from railway freight statistics. The results of this exercise, summarized in table 7, are obtained for seven major commercial crops, including wheat. Because information on the factory consumption of individual oil seeds is not available, estimation for the three selected seed crops, namely peanut, sesame, and soybean, has been undertaken for the crops as a group.

The results in table 7 fall very much in line with the impression conveyed by the spatial patterns of domestic trade flows rendered in figure 5. The expansion of interregional trade in farm products in prewar China was accompanied by the emergence of the national railway system as a primary trade carrier. By the mid-1930s, probably more than 60% of the trade volume in industrial crops, which constituted the core of the domestic agricultural trade, was railway-bound. In value terms railways' shares were even more pronounced because of the tendency for railways to absorb freight with higher unit values. Thus China's railways maintained a majority share in the carriage of long-distance farm trades, even though rice, which had lower unit value than the industrial crops, remained a largely water-borne trade commodity.

Railways and the Growth of Trade at Treaty Ports

While most of China's large treaty ports were initially chosen on account of their natural advantage in water accessibility, the later development of a national railway network which focused on these centers of commerce afforded strong stimulus to trading activities at the treaty ports. An indicator of this causal relationship is the synchronic pattern in the timing of railway construction and growth in the volume of outbound maritime trade recorded at the treaty ports. These statistics fail to fully describe the total contribution of railways to the treaty port economies, which apparently comprises total gains in trade as well as increases in local consumption. Total increment in the value of export trade from treaty ports is also understated using maritime records simply because railways became important carriers of export to non-local domestic markets. However, the treaty ports were by and large export-based economies, and the correlation between maritime trade and total trade was usually strong. On this basis a close examination of the maritime trade statistics can provide good insight into the

TABLE 7

THE SHARE BY RAILWAYS IN THE INTERREGIONAL
TRANSPORTATION OF MAJOR INDUSTRIAL
CROPS IN CHINA PROPER (1935)

Crop	Total Size of Trade ('000 metric tons)	Total Railway Tonnage ('000 metric tons)	Percent by Rail
Peanut Sesame Soybean	586^a	456^b	78
Cotton	455^c	290^d	64
Tobacco	133^e	86^f	65
Tea	52^g	35^h	67^i
Wheat^j	450	378	84

^{a}Total factory output of edible vegetable oils in China Proper was estimated at C$58.2 million in 1933 (Liu and Yeh, p. 428). Since available evidence indicated that consumer goods production was fairly stable between 1933 and 1935, and since the income elasticity of demand for food products is usually lower than consumer goods such as clothing, it can be safely assumed that output of food processing industries registered little variation between 1933 and 1935. (Note: Chang estimated that output of consumer goods, mainly cotton yarn and cotton cloth, actually declined slightly from the level of 1933=100 to 1935=93.5 [John K. Chang, p. 79t]. Output of cigarettes, as indicated by the Shanghai plant of the Nanyang Brothers Tobacco Company, also fell from the 1933 level of 116,474 cases to 109,887 cases in 1935 [ibid., p. 42t]. If these fragmentary data were representative of the consumer industries as a whole, the present estimation of interregional trade based on 1933 data for 1935 would result in upward biases.) The C$58.2 million figure is therefore taken to represent the level of output in 1935. Total value of seeds consumed by the industry was about 70% of output (Liu and Yeh, p. 444), or C$40.74 million. However, the seeds thus consumed included also rapeseed, cottonseed, and tea seed in addition to peanut, sesame, and soybean. Hence the 384,340 metric tons of oil seeds, derived by dividing C$40.74 million by the weighted average of seed prices, C$5.30/picul (weights being total seed value in oil extraction, see Liu and Yeh, p. 532), is an overestimation of the quantity of peanut, sesame, and soybean actually consumed by factories. As there is no way of telling what amount of the other seeds was consumed and since peanut was also consumed in kind, it was assumed here that this two quantities tend to cancel each other out. It is more likely, however, that 384,340 metric tons is a high estimate.

To this quantity the 1935 net exports of peanut, sesame, and soybean were added. These trade figures were obtained from Chinese Customs Reports for the year 1935 and came to a total of some 202,000 metric tons. The resulting total of 586,000 metric tons were taken as the size of domestic long-distance trade in peanut, sesame, and soybean in 1935. (Note: Manchuria produced only an insignificant amount of peanut and sesame. The Customs export statistics on these two crops therefore largely reflected the export from China Proper. On the other hand Manchuria was a major soybean exporter and accounted for the bulk of soybean export of China. In the present calculation soybean export from China Proper has been assumed to be nil.)

^{b}This is the sum of 72,000 metric tons of peanut, 116,000 metric tons of sesame, and 268,000 metric tons of soybean. The peanut figure, representing some 30% of the 1935 total railway tonnage in the crop, was mainly the quantity carried by the Lung-Hai, plus part of the tonnage handled by the Peking-Hankow and Tientsin-P'uk'ou lines. The sesame tonnage accounted for about 68% of the total carried by the railways in 1935. The Peking-Hankow and Tientsin-P'uk'ou railways handled the bulk of this traffic. These two lines also carried most of the estimated soybean tonnage, which was some 60% of the total handled by all the railway lines. (All railway freight statistics used refer to Yang Hsiang-nien, pp. 120-121).

TABLE 7--Continued

^{c}Consumption of raw cotton by modern mills in China Proper is estimated at 11 million piculs in 1935 (Liu and Yeh, p. 521). Deducting from this a net import amount of 1.9 million piculs (ibid.) gives 9.1 million piculs, or 455,000 metric tons, as raw cotton supplied from domestic sources.

^{d}This figure, accounting for some 38% of the total railway tonnage in raw cotton, was derived as the sum of the quantity carried by the Peking-Hankow, Lung-Hai, and Peking-Mukden lines. A minor proportion of the cotton carried by the Tientsin-P'uk'ou railway has also bee included.

^{e}This is the sum of tobacco leaves consumed by the cigarette manufacturing industry, 2.5 million piculs (Liu and Yeh, p. 560, note 62), and the amount of net export, 152,000 piculs (Ministry of Industry, Yen-yeh, pp. 133, 144).

^{f}This represents the aggregate tobacco tonnage carried on the Peking-Hankow, Tientsin-P'uk'ou, and Ch'ingtao-Chinan lines. This quantity accounted for 80% of the total tonnage carried by the railway system.

^{g}Factory output of tea in 1935 amounted to C$8.9 million (Liu and Yeh, p. 527), of which 79%, or C$7.03 million, may be considered the cost of raw materials (ibid., p. 444). At C$25.8 per picul, the amount of crude tea thus consumed came to about 272,480 piculs or 13,600 metric tons. In 1935 net export of Chinese tea totaled 38,140 metric tons (Chu Mei-yu, p. 146). Therefore total domestic trade in tea amounted to about 51,740 or 52,000 metric tons.

^{h}This is the total quantity carried by the Shanghai-Hangchou-Ningpo and Hupei-Hunan railways in 1935. It represented 33% of the total tonnage of tea carried by the railway system.

^{i}To the extent that steamers constituted a connecting mode of transport to the Hupei-Hunan railway for tea directly exporting from Hankow, the present estimate of railways' share in tea trade is upwardly biased.

^{j}See p. 65, notes 2, 3.

impact of rail transport development on the growth of China's major commercial and trade centers.

The maritime trade statistics recorded at the five largest treaty ports in China Proper during the period 1900-1930 are graphically portrayed in figures 6a through 6e. These line graphs trace out the total value of domestically-originated goods shipped out of the ports in three-year moving averages. The correlation between trade performance and railway development can be envisaged from the generally close timing of upsurges in trade values and the completion of railway lines as indicated by the vertical arrows. The following account provides some general description of the synchronic patterns of trade and railway development experienced at each of the treaty ports.

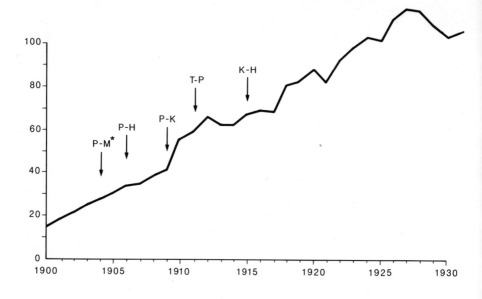

(a) Tientsin

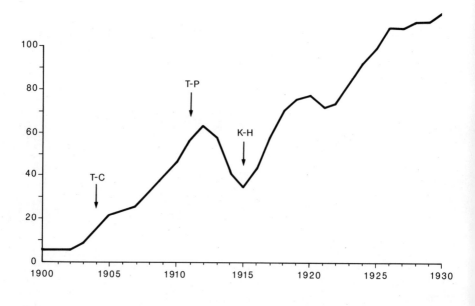

(b) Ch'ingtao

Fig. 6. Railways and maritime export trade at leading treaty ports, 1900-1930 (3-year moving average at 1926 prices, 1926=100)

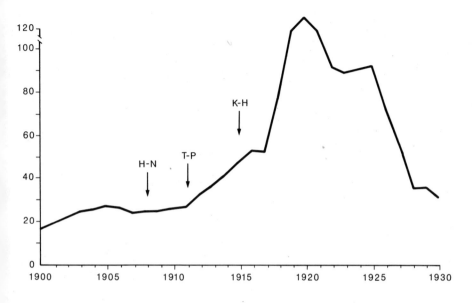

(c) Nanking

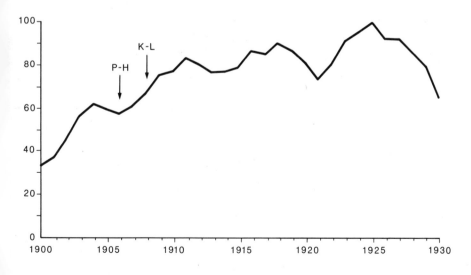

(d) Hankow

Figure 6 --Continued

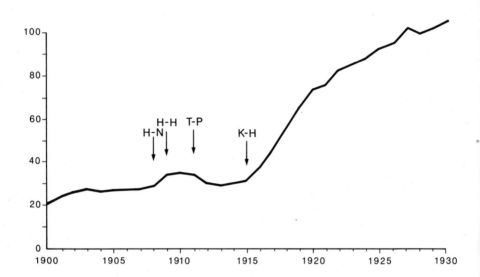

(e) Shanghai

SOURCES: Maritime Customs, <u>Reports and Returns of Trade at Treaty Ports</u>, various years from 1900 to 1930.

*Long vertical arrows indicate the first year in which the railway line was opened to full operation. The letter symbols for the trunk lines stand for, respectively (month and year in parentheses),

H-H: Shanghai-Hangchou railway (July 1909)
H-N: Shanghai-Nanking railway (March 1908)
K-H: K'aifeng-Hsüchou railway (section of Lung-Hai railway)(June 1915)
K-L: K'aifeng-Loyang railway (section of Lung-Hai railway)(Dec. 1908)
P-H: Peking-Hankow railway (March 1906)
P-K: Peking-Kalgan railway (August 1908)
P-M: Peking-Mukden railway (1904)
T-C: Ch'ingtao-Chinan railway (July 1904)
T-P: Tientsin-P'uk'ou railway (October 1911)

(1) Tientsin. The ascendancy of Tientsin as the trade and commercial capital of North China owed much to an expanded trade hinterland provided by five trunk railway lines. Between 1900 and 1912, the real value of Tientsin's export trade jumped more than

eightfold and its share in the national total of export-import trade rose from 1.3% to over 6%.[1] During this same period the Peking-Mukden, Peking-Hankow, Peking-Kalgan and Tientsin-P'uk'ou railways were successively completed, rendering Tientsin the most rail-accessible port in the entire country.

Two basic components may be identified in the surge of trade induced by railway development. One is the displacement of freight traffic from traditional modes of transport. The diversion of trade to railways confers economic benefits on the terminal port because of savings in terms of speed and regularity of delivery and lower risk of loss en route. But the economy at the regional or national level stood to gain from this change also because of the consequent reduction in the real cost of transport and the release of resources from traditional transports for employment in alternative production activities. For example, before segments of the Peking-Mukden railway were open to commercial traffic in the early 1890s, as many as 60,000 boatmen and carters were engaged in transport services between T'ungchou and Peking. By 1902, however, when the railway was thrown open to full operation, most of the T'ungchou-Peking trade had been captured by regular railway traffic,[2] this despite the relatively short-haul between T'ungchou and Peking.

A more important contributor to the rise of export trade, however, was the increased flow of rural exports stimulated by improved transport conditions. The concomitant rise in Tientsin's outbound maritime trade and the percentage of the port's inbound freight traffic carried by railways testified to this relationship. As early as 1906, rail-carried freight was accounting for some 48% of all goods arriving at Tientsin from the interior (table 8). The share of railways steadily increased through the years until 1922-24, when it peaked at 74%. In contrast water transport by traditional means declined from a high of 35% in 1906 to a low of less than 25% by 1922. In the post-1924 era, however, railways' margin of dominance was heavily eroded by military disturbances and the commandeering of rolling stock by the warlords. Even then, only in two years, namely 1926 and 1930, did the railways lose the lead to river junks in the percentage share of Tientsin's inbound freight traffic. Indeed, there is evidence that the railways regained their lead soon after political and economic conditions improved in

---

[1] Export trade statistics from C. Yang, et al., p. 92; trade-share figures from Chang Yu-kwei, p. 23, table 7.

[2] Maritime Customs, Returns of Trade at Treaty Ports for the Year 1902 (Shanghai, 1903), p. 54.

1931.[1] The resurgence of railways' domination was likely to have continued after 1931 with the official abolition of likin (transit duties), the general reduction in freight rates in the national railway system, and the prevalence of peace in North China until the late 1930s.

TABLE 8

SHARES IN GOODS ARRIVING AT TIENTSIN FROM INLAND

	Modes of Transport			
Year	Rail	Boat	Cart	Total
1906	48%	45%	7%	100%
1912	53	44	3	100
1914	55	43	4	100
1917	68	28	4	100
1921	72	26	2	100
1922	74	23	3	100
1923	74	23	3	100
1924	74	23	3	100
1925	66	31	3	100
1926	43	54	3	100
1927	50	46	4	100
1928	49	46	5	100
1929	54	42	4	100
1930	47	50	3	100

SOURCES: (Imperial) Maritime Customs, Returns of Trade at the Treaty Ports of China for 1906 (Shanghai, 1907), Part II, 1:160; idem, Decennial Report, 1912-1922 (Shanghai, 1923), 1:31; and idem, Decennial Report, 1923-1932 (Shanghai, 1933), 1:377.

(2) Ch'ingtao. The railway must be credited with Ch'ingtao's emergence as the second largest commercial port and industrial center in North China. From the beginning the port's economic and political promises were envisioned by foreigners to be closely intertwined with railway construction. This preconception precipitated Germany's demand for railway rights in Shantung and its colonization of Ch'ingtao in 1899. The construction of the Ch'ingtao-Chinan railway, completed in 1904, was as expected a big boost to Ch'ingtao's export trade (figure 3b). Part of the gain inevitably came from diverted commerce, which perhaps is most vividly reflected in the relative decline of Chefoo, a seaport located some 150 miles northeast of Ch'ingtao. Chefoo enjoyed an

---

[1] No data were available on modal shares of inbound freight since the Tientsin Native Customs was closed in 1931. Independent studies, however, revealed that railways carried 42% of the raw cotton arriving at the city during the first five months of 1931, as compared with only 19% in 1930 (Fong, p. 77)

unchallenged primacy in foreign trade in the Shantung peninsula until the Ch'ingtao-Chinan line was completed. Within four years from the opening of the railway to commercial traffic, Ch'ingtao's export multiplied by sixfold, while export trade at Chefoo declined slowly until the outbreak of the First World War.[1]

Apart from transient, war-related disruptions, such as Japan's intrusion into Shantung upon her declaration of war on Germany in 1914, Ch'ingtao's commercial position was well established soon after the construction of the Tientsin-P'uk'ou railway in 1911. The Tientsin-P'uk'ou line, together with the K'aifeng-Hsüchou railway completed in 1915, greatly stimulated the growth of Ch'ingtao by channelling rural farm exportables from Honan, northern Chiangsu and western Shantung to the port. Ch'ingtao, for example, remained the largest export center of peanuts in China before World War II, a success that can hardly be dissociated from the vital services of the great trunk lines.

(3) Nanking. Before the Tientsin-P'uk'ou railway accorded unrivalled commercial privilege to the city, Nanking was hardly of any national importance in the rank of trade centers. During the last decades of the Ch'ing dynasty, Nanking commanded some economic leverage as a provincial capital and treaty port. However, the port of Nanking was far from being one that could emulate the performance of such neighbouring counterparts as Wuhu and Chinchiang, not to mention the seemingly immutable preeminence of Shanghai.

The building of the Shanghai-Nanking railway contributed only marginally to Nanking's position as a national trade center. The reason is simply that the railway runs a parallel course with the lower Yangtze and added little to the port's trade hinterland. Export trade at Nanking exhibited a marked upturn in value only when the Tientsin-P'uk'ou railway suddenly threw open the door to Nanking's vast hinterland north of the Yangtze. When the P'uk'ou wharves opened in 1915, Nanking became an important entrepot capable of handling large volumes of sea and rail freight. As a result, steamer tonnage clearing the port between 1913 and 1921 increased from 79,420 to 552,224 tons, a magnitude of 600%.[2] From almost nothing in 1912, Nanking's direct trade with foreign countries grew to be in excess of 5 million <u>Haikwan</u> <u>taels</u> by 1920. Domestic

---

[1] Between 1903 and 1913 Chefoo's export trade dropped by 13%. During the World War I period, however, strengthened foreign demand for farm products considerably revived Chefoo's export trade, which grew by 50% in five years (see Maritime Customs, <u>Returns of Trade at Treaty Ports</u> [for years 1903, 1913, 1914, and 1918]).

[2] Maritime Customs, <u>Decennial Report for 1912-1921</u>, 1:359.

interport trade from Nanking also expanded drastically, so that total outbound trade of the port quintupled in less than a decade, from 4.6 million in 1912 to over 25 million Haikwan taels by the close of the 1910s.[1]

Despite adverse conditions such as the 1921 flood in central China, continued political strife, and tightness of the money market due to prolonged declines in the price of silver, Nanking's export trade progressed slowly throughout much of the 1920s. The Customs Reports attributed this mainly to the expanding services of railways which were actually capturing commerce from the steamers.[2] The precipitous dip in recorded maritime trade at Nanking in the late 1920s was primarily a result of the discontinuation of Customs records. The Maritime Customs was convinced that the bulk of outbound trade was carried by railways and therefore maritime trade registration could be abolished.[3] Rail transport thus appeared to have outvied water transport in a territory where shipping had long posed a formidable rival to overland conveyance.

(4) Hankow. Lying at the heart of the rich agricultural basins of the middle Yangtze and serving as the west-end terminal of ocean-going steamers plying on the big Yangtze River, Hankow secured a natual monopoly in the entreport trade of central China. Before the advent of railways, the magnificent system of natural waterways in Hunan and Hupei was responsible for carrying the bulk of the local farm exports to Hankow. The trade hinterland of Hankow, however, was geographically confined to areas easily accessible to river channels. Regions north of the Han River system, with perhaps the exception of the Wei River basin, had little trade relationship with Hankow simply because of the high cost of land transport.

The construction of the Peking-Hankow railway and the K'aifeng-Loyang section of the Lung-Hai railway significantly modified the spatial structure of Hankow's trade hinterland. While agricultural products from the larger part of Hupei and Hunan still found their way to Hankow via rivers and streams, a conspicuous flow of rural products from Honan, Hopei, Shanhsi, Shenhsi, and as far as Mongolia, reached Hankow via the railway. The new trade with the

---

[1] Ibid., p. 427.

[2] Total steamer tonnage cleared the port was 9.7 million tons in 1922, but it dropped to a little over 9.1 million tons by 1931 (see Maritime Customs, Decennial Report for 1922-1931, 1:620).

[3] The Maritime Customs had the opinion that "even when likin was levied on rail-borne cargo, merchants already found it cheaper to send certain goods by land than by water; since the abolition of likin in 1931, the land route has become even more attractive" (ibid., p. 619).

Northwest and North China added substantially to Hankow's position as an export center and enabled the Yangtze port to compete with such seaports as Tientsin, Ch'ingtao, or even Shanghai.

Of course, not all the rail-borne freight arriving at Hankow from the north was "new" in that it was generated by the railways. Sesame seed from southwestern Honan and native exports or re-exports from Hsian in the Wei River basin, for example, were goods diverted to the railway route from the former water course along the Han River.[1] One major reason for this diversion of trade is the saving in time of shipment, since the railways could achieve in one or two days what normally took the river junk over 20 days to accomplish.[2] Reduction in the risk of cargo loss en route was another important reason for railways' successful capturing of the major share in the conveyance of the more valuable commercial crops and livestock products. Hides and herbs from the Northwest, cotton and tobacco from Shenhsi, Kansu, and western Honan, and even tea from the hilly terrains of northeastern Hunan were good examples of railway's successful rivalry with traditional river transport.

It goes without saying that the big improvement in Hankow's trade position since the late 1900s was linked to the vital service of railways, not because of railways' successful capture of existing trade, but because of the expansion of agricultural commercialization in regions the railway line had come into presence. The increased production of sesame seed, cotton, and peanut along Honan's railways contributed much to Hankow's export trade, as was the increased exportation of cereal crops along the Peking-Suiyuan railway.[3] When civil war escalated in the 1920s and the service of the Peking-Hankow railway, among others, was severely impaired, export trade at Hankow stagnated correspondingly. In 1925, the Lung-Hai railway was finally extended to Haichou where regular steamer service provided transport links to Ch'ingtao and Shanghai. As a result the bulk of farm exports from Honan and Shenhsi became diverted to an eastbound route, and export trade at Hankow declined perceptibly. The prolonged depreciation of silver versus the Hankow tael helped stimulate exports at Hankow but failed to revive the port's deteriorating trade position.[4] It was not until the completion

---

[1] See Ts'ui-hung Liu, pp. 42-43, 178-79.

[2] Ibid., p. 23.

[3] For further discussion of empirical evidence in this regard see chapter V below.

[4] Maritime Customs, Decennial Report for 1922-1931, 1:554-56.

of the Hangyang-Yüehchiang section of the Hankow-Canton railway in 1936 that fresh stimulus for export growth was again experienced at Hankow. By then, however, impending war with Japan had begun to erode much of the hope of a complete revival in the trade performance of the great Yangtze port.

(6) Shanghai. Shanghai owed its early prominence in export-import trade to the great hinterland of the Yangtze basins where the bulk of China's interregional trade occurred. Railway construction which conferred special commercial advantage on such Yangtze ports as Hankow and Nanking since the turn of the present century actually helped strengthen Shanghai's role in domestic and international trade. This explains at least partly why Shanghai's share in the national total of native exports to foreign and domestic destinations improved, though slightly, from about 24% in the early 1890s to 28% in the early 1930s.[1]

As China's largest commercial, industrial and trade center, Shanghai was the natural focal point of modern communication and transport networks. The Shanghai-Nanking and Shanghai-Hangchou railways, both completed in the late 1900s, considerably enhanced the political and commercial importance of the city, but mainly by attracting passenger traffic rather than freight. The basic reason was simply that both trunk lines failed to extend beyond Shanghai's immediate hinterland, the Yangtze delta region, which had long been served by a magnificent system of waterways. The building of the Tientsin-P'uk'ou railway, though it provided the Yantze delta with a direct transport link to most of North China, actually did Shanghai a disservice by encouraging export trade at Nanking. Consequently Shanghai's direct trade with foreign countries diminished by 38% between 1911 and 1914, from 86 million to 53 million Haikwan taels in 1926 prices.[2]

It is generally difficult to discern the relationship between railway development and export trade growth at Shanghai from Customs trade statistics. For one thing, a sizable portion of inflowing commerce was ultimately absorbed by Shanghai's industries as material inputs. For another, Shanghai's exports comprised not only raw and semi-processed agricultural commodities but a large amount of manufactured products as well. In fact, it is more appropriate to relate railway development to industrial growth at Shanghai and regard

---

[1] Based on statistics in Maritime Customs, Returns of Trade at Treaty Ports (various years).

[2] See Maritime Customs, Decennial Report for 1912-21, 1:427 and Decennial Report for 1902-1911, 1:xvii.

outflowing maritime commerce from the port as lagged indicators of local industrial expansion. The insight one gets from this perspective is the growing importance of railways in the carriage of domestically originated industrial crops and certain foodgrains, as mentioned earlier in this chapter. With the abolition of likin and expansion of intermodal freight links in the early 1930s, Shanghai stood to gain further from growth in rail-borne freight, even though these developments turned out to have adversely affected such interior ports as Nanking and Hankow.

Summary

The development of long-distance agricultural trade in China during the first few decades of this century summarized the salient aspects of an ongoing process of agricultural transformation. New forces of demand generated by the rise of industrialization, accelerated urbanization, and growth of international trade provided opportunities for the profitable expansion of the agricultural surplus. The development of rail transport, on the other hand, considerably reduced the cost of supplying farm products to the cities and in effect reinforced the growth of the demand for agricultural output. This impact of the railways on the peasant economy was most clearly discerned in North China, where geographical segregation between coastal areas of demand and interior areas of supply was both a matter of long physical distances and one of the lack of navigable waterways.

While China's long-distance agricultural commerce was dominated by a few luxurious items of export besides rice before the 1900s, considerable expansion and diversification in the composition of this trade occurred with the development of rail transport after the turn of the century. By the mid-1930s, railways had superseded the river junk as the major carrier of long-distance agricultural trade (by value). Railways were particularly important in the conveyance of industrial crops, of which probably more than 60% (by volume) were rail-borne in 1935.

The synchronic pattern of growth in the export trade of leading treaty ports and the construction of railway lines also testified to railways' impact on the development of domestic commerce. An examination of Maritime Customs statistics reveals that the construction of railway lines was usually accompanied by upsurges in the export trade of affected treaty ports. To the extent that the economic growth of China's treaty port economies was reflected in the growth of export trade, an intimate relationship was evident between railway development and the growth of China's modern sector.

CHAPTER IV

THE IMPACT OF RAILWAYS ON AGRICULTURAL PRODUCTION

Introduction

The preceding chapter has explored the relationship between railway extension and the development of long-distance agricultural trade up to the mid-1930s. It has left unanswered the connection between rail transport development and the growth of agricultural output, though the argument presented often assumed the importance of this relationship in the observed process of agricultural commerce expansion. This chapter and the next take on this latter issue by examining available empirical evidence, with the use of statistical analysis where appropriate and the support of qualitative information.

The inquiry is conducted within the context of three hypotheses. The first hypothesis states that reduction in the transport cost of non-local agricultural marketing increased farm output and factor input utilization by peasants in prewar China. The second hypothesis postulates that railway development in prewar China reduced transport cost in long-distance agricultural marketing and therefore stimulated farm production. Implicit in both hypotheses is the enhanced market participation of the agricultural producer, which furnishes the link between agricultural growth and the expansion of long-distance farm trade. This commercialization process and its relation to railway development is brought, in the next chapter, into the inquisitional focus of a third hypothesis. This last hypothesis asserts that railway development provided a strong drive for the expansion of commercial cropping and affected the pattern of agricultural commercialization in prewar China.

Because of the lack of cross-section time-series data on farm production, no direct analytical investigation of the effects of railways on agricultural production is attempted. The formulation of the first two hypotheses are thus geared to a methodology which permits reasonable inference from the analysis of cross-section data alone. However, the availability of a relatively rich body of cross-section time-series data on farmland values makes possible a close examination of synchronic changes in land productivity and transport conditions. This analysis is carried out in a later section of this chapter.

Market Accessibility and Agricultural Productivity: Hypotheses

Improvement in transport efficiency has commonly been recognized as an important factor in the agricultural development of low-income countries.[1] Higher farmstead prices of agricultural products, which accompany improvement in transport and marketing facilities, encourage specialization in cash crop cultivation and enhance farm productivity based on mobilization of previously unutilized and/or underutilized resources.

That farm production responds positively to improvement in transport conditions, such as those deriving from the construction of railways, describes the general propensity of peasants to respond rationally to changes in the relative price of farm outputs. Empirical studies of low-income agricultures confirm the existence of significant and positive price responsiveness of agricultural output of an order of magnitude quite comparable to that observed in the agriculture of high-income areas.[2] Though many of these studies are concerned with single crop response to relative price changes and not the price responsiveness of total farm production,[3] they provide sound evidence in corroboration of the working of the profit motive in farm operations even in traditional economies.[4] As

---

[1] See, for example, J. C. Abbott, "The Role of Marketing in the Development of Backward Agricultural Economies," The Journal of Farm Economics 44 (May 1962): 349-62; John W. Mellor, The Economics of Agricultural Development (Ithaca, N.Y.: Cornell University Press, 1966), pp. 328-43; Colin Clark and Margaret Haswell, The Economics of Subsistence Agriculture (New York: St. Martin's Press, 1967), pp. 179-99; and Margaret Haswell, Tropical Farm Economics (London: Longmans, 1973), pp. 115-29.

[2] For a summary presentation of the econometric evidence, see Jere R. Behrman, Supply Response in Underdeveloped Agriculture (Amsterdam: North-Holland, 1968), pp. 15-18; see also Behrman's own results on Thai rice-planting (ibid., pp. 316-33). Such studies have tended to show price elasticities (i.e., percent change in output effected by 1% change in relative price) ranging from about 0.1 to 0.5 or 0.6 for major foodgrains, and 0.5 to 1.1 for nonfood crops.

[3] Very few estimates of the price elasticity of aggregate agricultural production have been made. For one study in this area see Robert W. Herdt, "A Disaggregate Approach to Aggregate Supply," American Journal of Agricultural Economics no. 4 (November 1970):512-20. The price elasticity of aggregate agricultural production is lower than for individual crops, since substitution between crops often accounts for an important portion of the response of a single crop to changes in its relative prices. Even so, substantial gains in total output value can be realized as more remunerative (higher value per unit weight) cash crops are substituted for subsistence food staples in the event of a general rise in farm prices and/or relative rises in the prices of cash crops.

[4] The most noted proponent of this view of the "optimizing peasant" is Professor T. W. Schultz, who has emphasized the innate tendency of agricultural cultivators to respond positively to economic incentives within their perceived opportunity sets and to allocate resources efficiently as they learn through

technological advance in the transport sector results in transport savings and corresponding rises in farm prices, rational responses to profit incentives on the part of farm operators will work toward significant gains in the aggregate value of farm production and farm income.[1]

However, new opportunities afforded agricultural producers by transport advances are not limited to changes in the relative price of agricultural outputs. Market expansion and improvement in the efficiency of market performance are among the most significant accompaniments of transport development. Not only do these changes in the economic environment of exchange help reduce the erratic nature of agricultural price fluctuations and thus increase the price

experience (see Schultz, Transforming Traditional Agriculture [New Haven: Yale University Press, 1964]). The weight of research evidence supports this view. For examples of qualitative accounts of the price responsiveness of Chinese peasants in the prewar period, see chapter V, p. 100.

[1] The economic mechanism involved may be formulated mathematically, more or less along the lines of a Thünen-type model. Consider a farming community G located some distance from a central market M, to which the output of G (assumed for simplicity to consist of one single homogeneous product) is shipped for sale. Farm price at G, p_c, is the ruling price at M (which is given) net of the unit cost of transport from G to M. Thus p_c is a decreasing function of the transport rate t:

$$p_c = p_c(t) \quad \text{and} \quad (dp_c/dt) < 0$$

Let the cost function of farm production at G be of the form

$$C = F + a\,(Q/L)^b$$

where $a > 0$, $b > 1$, and C, F, (Q/L) denote, respectively, cost per acre, fixed cost per acre, and output (Q) per acre. Land of a homogeneous quality at G is fixed in supply such that $L = \underline{L}$. In a system of owner-operator farms the profit function of farm production at G is thus

$$\Pi = p_c(t)Q - \underline{L}\,F - a\,\underline{L}\,(Q/\underline{L})^b$$

The profit-maximizing output Q* is given by solving

$$d\Pi/dQ = 0$$

or $\quad Q^* = \{[p_c(t)/ab]\,\exp\,(1/b-1)\}\,\underline{L}$

where exp is the exponential function (i.e., a exp (x) = a^x). Differentiating Q* with respect to t, we get

$$(dQ^*/dt) = \{[p_c(t)/ab]\,\exp\,(2-b)\}(\underline{L}/b-1)(dp_c/dt) < 0$$

In other words equilibrium output increases as the transport rate decreases and vice versa. Note that if substitution of higher-valued crops for lower-valued crops occurred, higher profit would result in spite of no further change in Q*.

elasticity of output supply,[1] they also contribute to increased market perfection and hence to farm productivity through increased efficiency in the allocation of productive resources.[2]

It is thus plausible that structural changes in transport systems, by generating and modifying the pattern of geographical differentiation in the market accessibility of agricultural producers, should have left strong impressions on the pattern of farm productivity. A spatial equilibrium would be achieved, in which more accessible areas (in terms of the relative economic distance to regional markets) received higher incentives to produce and specialize in higher-valued commercial crops because of higher farm prices and improved market efficiency.[3] In prewar China, urban and industrial expansion and the upsurge in export demand occurred fortuitously at a time when population pressure had begun to strain beyond the capacity of a traditional production technology for maintaining stable per capita consumption levels.[4] The economic impulse to increase specialization and commercialization, bred from an innate drive away from the course of deterioration in the standard of rural living, was afforded fresh impetus from the demand side.

However, the spatial incidence of the impact of changing market demand was not a uniform process. The evolution of a national network of rail transport conditions accentuated the areal differential growth in farm productivity; but instead of encouraging

---

[1] Large fluctuations in agricultural prices suppress producer incentives to produce for the market since it elevates the risk of having to buy subsistence needs at very unfavourable price ratios should the cultivator choose to produce commercial crops at the sacrifice of food staples. Reduction in price fluctuations, both on a seasonal and year-to-year basis, would reduce this risk and encourage the increased reallocation of resources from subsistence to market production. This not only would increase the supply of commercial crops, but would improve the market prices of subsistence crops (due to decreased aggregate supply) and their supply elasticity as well (see Mellor, pp. 200-1). There is no lack of evidence in prewar China of the actual and potential impacts of transport improvement on local price fluctuations and productivity growth. For one example that deals particularly with the contribution of rail transport, see Bank of China, Economic Research Unit, Mi (Rice), p. 149.

[2] More discussion and explicit testing of this statement will be found in a later section of this chapter.

[3] Conditions conducive to improved market efficiency include damped local price fluctuations, increased market competitiveness, and a faster and more effective interaction between spatial markets due to declining time-distances.

[4] Many scholars have argued that by the late 19th and early 20th century, Chinese agriculture had reached the point of sharply diminishing returns to labor because of a stagnant technology in agricultural production and the growing scarcity of productive land resources (see, for example, Dernberger, pp. 24-26; Perkins, Agricultural Development, pp. 26-29; and Mark Elvin, The Pattern of the Chinese Past (Stanford: Stanford University Press, 1973), p. 312).

a polarization of condition in rural areas, the development of rail transport stimulated the specialization of commercial cropping in rail-accessible areas as well as the market production of subsistence food staples in less accessible regions.[1] In other words, increased specialization in the production of commercial crops in areas of improved transport conditions also enhanced the comparative advantage of increased production and marketing of subsistence crops elsewhere. The consequence was a spatial continuum of farm productivity levels in accord with continuous changes in accessibility over economic space.

In summary, rational response on the part of agricultural producers to market forces suggests a definite relationship between the relative cost of agricultural marketing and the spatial variation in farm productivity. The postulated existence of this relationship in prewar agrarian China leads to the following hypothesis:

Hypothesis I: Farm productivity, defined as total farm income from agricultural production, would be positively related to the accessibility of (or negatively related to the total cost of transportation from) the farm to the nearest (in economic distance) regional market. It is hypothesized that increases in accessibility tended to induce both higher levels of factor inputs and a more efficient utilization of resources in farm production.

Empirical Testing of Hypothesis I[2]

The empirical testing of hypothesis I is given in two parts. The first part deals with the relationship between accessibility and, respectively, farm output, total inputs, and disaggregated input levels. The second part of the test will examine the impact of accessibility changes on resource allocation efficienty.

Farm Productivity and Market Accessibility

The empirical data used in the present test are based mainly on the original work of J. L. Buck,[3] the only large-scale study on the production aspects of prewar rural China. The primary data

---

[1] For more discussion of this point see chapter V, p. 114.

[2] Analyses presented in this and the following sections draw heavily on Ernest P. Liang, "Market Accessibility and Agricultural Development in Prewar China," Economic Development and Cultural Change 29 (September 1981).

[3] J. L. Buck, Land Utilization in China and idem, Land Utilization in China Statistical Volume (Chicago: University of Chicago Press, 1937), hereafter abbreviated as LUSV. Buck's study was based on extensive field surveys carried out mainly between 1929 and 1932 on 22 agriculturally productive provinces in China Proper. It covered 16,786 farms in a total of 168 localities, though not all production data were available in all of the localities studied.

input in this study comprises cross-section series on average farm output, seven factor inputs, and an independently estimated index of accessibility. Consideration of data consistency allows only 141 localities (hsien or counties) to be chosen for this study, of which 63 are in the Wheat Region and 78 in the Rice Region.[1] The locations of the selected localities are shown on figure 7 and the agricultural regionalization on figure 8.

The estimation of farm output and accessibility is presented in some details in appendixes A and B respectively. The farm output figures represent the expected value of the aggregate agricultural output of an average farm in a locality. Output is expressed in terms of kilogram equivalents of the most commonly consumed grain in the local area,[2] which for convenience is taken to be wheat and rice in the Wheat and Rice Regions respectively.[3] Presumably each output figure designates the "expected" or "normal" output of farms in a locality so that the level of use of factor inputs was independent of random disturbances arising from natural or man-made factors which might have interferred with the pure effect of accessibility on input levels and hence on farm output.[4] Accessibility, on the other hand, is measured by the reciprocal of the total cost of transporting a ton of farm products from the farm to the regional market of sale. In effect accessibility is equal to the negative value of the total cost of transfer when both values are expressed in logarithmic terms.[5]

---

[1] The agricultural regions in this study are the ones defined by Buck (see Buck, Land Utilization, p. 25).

[2] If local prices of the most commonly consumed grain are taken as proxies of local general price levels, the output figures can be interpreted as real output values indicative of the real level of potential farm income. Since the actual level of prices was likely to consist of prices of imported goods as well as traded and non-traded commodities of local origin, the use of a single crop price instead of the general price level (e.g., some weighted average of all commodity prices) in fact tended to dampen the spatial variation in farm output values.

[3] Wheat, which supplied on the average 24% of the total calorie intake from staple crops in a North China farm household, was the most commonly consumed grain in the Wheat Region. Likewise rice topped the list of foodgrains in the Rice Region by accounting for nearly 70% of the calorie consumption of an average farm household in the region (see Buck, Land Utilization, p. 413). Though local exceptions to such identifications apparently existed, they are unlikely to affect the results of the present study appreciably. The use of wheat and rice prices as common yardsticks of measurement not only allows consistency in estimation, it is indeed essential if the spatial variation in farm prices is the object of inquiry, as in the present study.

[4] See the discussion in pp. 70-71 below.

[5] Let A and T be the index of accessibility and the total cost of transfer, respectively, then by definition

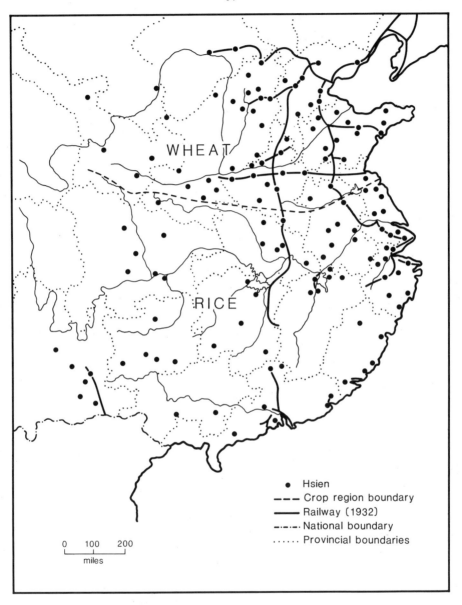

Fig. 7. Location of sample hsien (counties)

Fig. 8. Agricultural regions of China (after J. L. Buck)

The empirical relationship between market accessibility and farm output can be estimated by fitting equation (1) to empirical data with the ordinary least squares (OLS) method:[1]

$$\ln Y^* = \ln c + \gamma \ln A \qquad (1)$$

where ln = natural logarithm; Y^* = expected average farm output (1933); γ (with an anticipated positive sign) = elasticity of farm output with respect to accessibility;[2] A = measure of accessibility; and c = a constant parameter to be estimated. The regression results for the two crop-regions are displayed in table 9 and are both statistically significant at the 1% level of confidence. The γ-values as estimated suggest that each percentage point increase in the value of accessibility (i.e., each percentage point decline in total transfer cost to market) resulted in about 0.3% increase in average farm output. The magnitude of increase was somewhat higher in North (Wheat Region) than in South China (Rice Region).[3] These regression results are therefore strongly in support of the existence of a positive relationship between market accessibility and farm productivity in prewar rural China.

$$A = 1 / T$$

Taking logarithm of both sides we have

$$\ln A = \ln (1/T)$$
$$= - \ln T$$

i.e., the logarithm of accessibility is equal to the negative of the logarithm of the total cost of transfer.

[1] A pooled estimation of the production function of the form of equation (2') (see below) for the entire sample with a dummy variable for regions exhibited significant difference in the production relation in the Wheat and Rice Regions. For this reason all analyses in this study will be conducted separately for the two crop-regions.

[2] In mathematical terms, the elasticity of a variable Y with respect to another variable X is given by

$$\varepsilon_{yx} = d\ln Y/d\ln X = (dY/Y)/(dX/X)$$

[3] Possible errors in the measurement of accessibility, due to the mis-identification, mis-specification, or omission of markets and/or shipment routes, would result in downwardly biased elasticity estimates. If the vector of measurement error has variance δ_v and the sample variance of the true values of accessibility is $\delta_{\bar{x}}$, then the asymptotic bias of the estimate of the true slope γ is

$$\text{plim } \gamma^* = \gamma / [1 + (\delta_v^2 / \delta_{\bar{x}}^2)]$$

where plim is probability limit (see J. Johnston, Econometric Method, 2nd ed. [New York: McGraw Hill, 1972], p. 282).

TABLE 9

REGRESSIONS RELATING FARM OUTPUT TO ACCESSIBILITY

Region	Constant	Slope(γ)	F-Ratio	R^2	Sample Size	(Geometric) Mean Output	Mean Accs.
Wheat	8.553	0.317	76.27*	0.556	63	1915	0.046[a]
Rice	8.831	0.278	51.70*	0.405	78	3285	0.079[a]

SOURCE: See appendixes A and B.

*Significant at 0.01 level.

[a]These mean accessibility indices correspond to a total transfer cost of C$21.85 and C$12.68 in the Wheat and Rice Regions respectively. Since the accessibility index is the reciprocal of total transfer cost, it is unbounded above but has a lower bound of zero (least accessible).

Improved transport conditions, and thus better accessibility, tended to raise prices to producers who may respond with a significant added input of resources in production for the market and significant increases in overall output. This factor-augmenting effect of increased accessibility can be detected in the following manner.

Assume the stochastic Cobb-Douglas production function of the form[1]

$$\ln Y_j = \ln k + \sum_i \alpha_i \ln X_{ij} + u_j \qquad (2)$$

where ln = natural logarithm, j = j-th locality, Y = actual realized farm output for the year under concern, X_i = the i-th factor input, α_i = the elasticity of output with respect to input X_i, k = a constant parameter, u = a disturbance term assumed to be randomly distributed among localities. The random variable u represents collectively the effect on realized output of uncontrolled factors such as weather, pest infestation, other "acts of nature," and even the temporary outburst of warfares. Equation (2) may be estimated by the ordinary least-square method, but the estimated α_i coefficients would be

---

[1]Owing to its basic consistency with the established body of economic theory and computational simplicity, the Cobb-Douglas function is a popular specification of production relations (see E. O. Heady and J. L. Dillon, Agricultural Production Functions [Ames: Iowa State University Press, 1961]). The choice of a stochastic function here is designed to get around a major theoretical difficulty in tests for allocative efficiency using the function (see below).

inconsistent and biased estimators of the true α_i values.[1] If however peasants maximized <u>expected</u> profit corresponding to a <u>planned</u> level of output and if u consisted of exogenous shocks unanticipated by cultivators at the time of resource allocation, the production function (2) can be rewritten as[2]

$$\ln Y^* = \ln k + \sum_i \alpha_i \ln X_i \qquad (2')$$

where Y^* = expected output (equal in mathematical terms to the expected value of the random variable Y); the other variables are as defined above. Ordinary least-square estimation of (2') would then yield consistent and unbiased estimators of α_i (i = 1, . . . , n).

Equation (2') was estimated with the estimated output series and seven factor input series.[3] The results, given in table 10, are ordinary least-square estimations of (2') with different versions of the land variable entered for comparison and interpretation purposes. The estimates of the production function coefficients show generally an excellent fit, and estimated coefficients significantly different from zero in all cases except one. The insignificant irrigated land input coefficient in the Wheat Region, however, essentially reflected the predominance of the dry-farming system in North China.[4] The rice culture of South China, on the other hand, revealed itself distinctly in the high elasticity of the irrigated land input in the Rice Region. The regression results also bear out a common characteristic of traditional agricultures: that land (particularly land measured in terms of cropped area in the present case) and labor constituted

---

[1] This is known as simultaneous equation bias. See, for example, P. A. Yotopolous and J. B. Nugent, Economics of Development (New York: Harper & Row), p. 84.

[2] See Irving Hock, "Estimation of Production Function Parameters Combining Time-Series and Cross-Section Data," Econometrica 30 (January 1962):34-53. For an alternative specification of the model, see A. Zellner et al., "Specification and Estimation of Cobb-Douglas Production Function Models," Econometrica 34 (October 1966):784-95.

[3] An eighth variable, farm implement, has been dropped from the production function analysis due to its high correlation (R^2 = 0.834) with the farm building variable (to avoid multicollinearity problems). The farm implement variable was estimated by assigning weights to various implement items according to their relative prices. Since such weights were specified in ordinal scales and not in money value terms, the implement variable is different from the building variable in the unit of measurement, and the two cannot be combined to form a composite index of farm capital values.

[4] See Buck, Land Utilization, pp. 186-88.

TABLE 10

ESTIMATION OF EQ. (2'): COBB-DOUGLAS PRODUCTION FUNCTION ANALYSIS OF CHINESE AGRICULTURE: 1933

Region	Regression No.	Constant	Variables							Sum of Elasticities	R^2
			X_1	X_2	X_3	X_4	X_5	X_6	X_7		
Wheat	2'a	6.060	0.383** (0.107)	0.261** (0.109)	...	...	...	0.205** (0.061)	0.108** (0.024)	0.957	0.887
	2'b	6.177	0.348** (0.112)	...	0.295** (0.076)	...	...	0.164** (0.058)	0.101** (0.023)	0.908	0.897
	2'c	6.130	0.357** (0.109)	...	...	0.076 (0.045)	0.161* (0.070)	0.210** (0.058)	0.107** (0.023)	0.911	0.888
Rice	2'd	6.750	0.216** (0.064)	0.387** (0.056)	...	...	...	0.085* (0.038)	0.189** (0.036)	0.877	0.869
	2'e	6.865	0.174** (0.065)	...	0.442** (0.061)	...	...	0.082** (0.037)	0.134** (0.037)	0.832	0.874
	2'f	6.988	0.185** (0.063)	...	...	0.423** (0.052)	0.036** (0.013)	0.125** (0.033)	0.107** (0.036)	0.876	0.889

SOURCES: See appendix for sources and derivation of output. X_1 = labor (average man-equivalent per farm; a man-equivalent is defined by Buck as one adult male unit performing twelve months of farm work), from LUSV, p. 286; X_2 = cultivated land (hectares in crops), from LUSV, p. 286; X_3 = cropped area (hectares of crop raised in one year on the same piece of land), from LUSV, p. 286; X_4 = percent cropped hectares irrigated ((LUSV, p. 214] xX_3); X_5 = cropped hectares not irrigated ($X_3 - X_4$); X_6 = metric tons of fertilizers applied (derived from Land Utilization, p. 259, table 13); X_7 = value of farm buildings in C\$, derived from LUSV, pp. 357-62.

* Result statistically significant at 0.05 level of confidence.

** Result statistically significant at 0.01 level of confidence.

the principal input factors in prewar Chinese agriculture.[1] Overall, the specified inputs were able to account for more than 85% of the variance in output (as indicated by the values of R^2), implying that accessibility and the level of aggregate factor inputs were positively correlated.[2]

---

[1] If constant returns to scale, perfect competition, and profit maximization are assumed, the estimated input elasticities would represent the relative shares of the factors of production in total output (see Yotopolous and Nugent, p. 68). Thus from the results in table 10, land and labor together invariably accounted for more than 60% of total output in both crop-regions. It should however be noted that for two particular reasons, the estimated share of labor, or labor elasticity, is an underestimate. First, Buck's labor unit, man-equivalent, measures labor on farm-works only. The present output estimates, on the other hand, included also animal products which Buck considered as output of subsidiary, non-farm-works. Since animal output was likely to increase its share in total output as accessibility improved (Buck recognized animal husbandry as an increasingly important cash-income source [see Land Utilization, p. 298], total share of labor in output should be larger than the estimated labor time. Again, if higher output and better accessibility implied less idle time for laborers, which is the case with positive marginal productivity of labor, the estimated labor share would be underestimated. In short, both sources of error indicate the fact that a more carefully defined labor measure in service-flow units would increase the estimated share of labor (or labor elasticity) in farm production.

[2] To what extent could output growth be attributed to growth in total productivity, or technological change, associated with accessibility improvement? Existing qualitative evidence tends to suggest the possibility of technical change as a source of output growth in prewar Chinese agriculture. For example, the introduction of better quality and higher yielding imported species of crops such as cotton, tobacco and peanuts was often confined to more market accessible areas, particularly those along railway routes (see, for example, Chen Han-seng, pp. 6-7, 19; Li and Chang, CKNY, 2:159-60, 2:164; and Nanking University, Department of Agricultural Economics, Yu, Ngo, Yuan, Kan shih-sheng chih mien-ch'an yun-hsiao, p. 33). Changes in technical efficiency was also likely to be embodied in better capital equipment, better educated and more knowledgeable farm operators, and more efficient organization of production within and among farms in areas served by better transport facilities. The measurement of embodied, or non-neutral (in the Hicksian sense) technical change cannot be pursued in this study due to the lack of data. But the role of disembodied technical change can be roughly assessed by means of the geometric index of productivity as first suggested by Robert Solow ("Technical Change and the Aggregate Production Function," Rev. Econ. Stat. 39 [August 1957]:312-30). The approach is simply to treat technical change as a residual of real output growth unaccounted for by increases in real inputs, i.e.,

$$\Delta K/K = \Delta Y^*/Y^* - [\sum_i \alpha_i (\Delta X_i/X_i)]$$

where Δ is the difference operator; K is a measure of the "level of technology;" and the other variables are as defined in eq. (2') above. Taking technical change as a result of accessibility improvement (instead of the elapses of time), the formula was applied to the present data which have been arbitrarily divided into two groups at the geometric mean of the index of accessibility. Expressed as shares in output growth, technical change was thus found to be about 7% and 4% in the Wheat and Rice Regions respectively. These figures, of course, failed to accord an important role to technical change in output growth, especially when the possibility of omitting relevant factor inputs arises. It should be noted, however, that the smallness of the sample size could have resulted in an underestimated share of technical change.

The disaggregated effects of accessibility change on individual factor input use can be estimated by the following system of regression equations:

$$\ln X_i = \ln c_i + \sum_i \beta_i \ln A \qquad (i = 1, \ldots, 7) \qquad (3)$$

where X_i and A denote the i-th factor input and accessibility respectively; c_i is a constant; and β_i is the elasticity of input X_i with respect to accessibility. Table 11 summarizes the estimation results. The obvious conclusion one can draw from these regression results is the consistent tendency for factor inputs to rise with improved accessibility. Again, apart from the irrigated land variable in the Wheat Region, all estimated coefficients are significantly different from zero at the 0.01 level of confidence.[1] The lack of association between accessibility and irrigated land area in the Wheat Region is attributable, at least in part, to North China's prevalent system of dry-farming, as is evidenced by the high elasticity of non-irrigated land input. An additional contributing factor is the fact that railways had improved the accessibility of many localities away from waterways which served both as low-cost transport channels (as compared with other traditional modes of transport) and sources of irrigation water. Perhaps more noteworthy is the generally more elastic response of cropped (multiple cropping) area than cultivated area to accessibility changes in both crop regions. The differential performance of these two land variables, however, demonstrated the rational response of agricultural producers to profit incentives under land resource constraints.

The economic mechanism involved may be simply stated. Better trade opportunities attendant upon improved accessibility stimulated production and increased the derived demand for all production factors, including land. Given the scarcity of arable land in densely-populated agrarian China, land supply may be augmented by resorting to expansion at the extensive and/or intensive margins of cultivation. As farm prices improved with better transport conditions, the effort at expanding land resources under production also intensified. This relationship is borne out by the regression results exhibited in table 12. Part (a) of table 12 shows an empirical relationship between the percentage of farmland under crops

---

[1] The estimated labor coefficient is an underestimate for reasons put forth in p. 73, note 1 above. The farm capital (X_7) coefficient, on the other hand, is probably an overestimate, since fixed capital assets have been valued at local prices without adjustment for regional variations in general price levels which might closely correspond to variations in accessibility.

TABLE 11

REGRESSIONS RELATING INPUT LEVELS TO ACCESSIBILITY

Dependent Variable	Wheat Region				Rice Region			
	Constant	Slope	F-Ratio	R^2	Constant	Slope	F-Ratio	R^2
X_1: Labor	1.384	0.259	33.19**	0.352	1.186	0.183	14.11**	0.157
X_2: Cultivated land (crop area)	1.589	0.325	18.05**	0.228	0.886	0.334	44.52**	0.369
X_3: Cropped area	2.158	0.426	34.92**	0.364	1.441	0.350	45.66**	0.375
X_4: Irrigated cropped area	-0.249	0.074	3.73	0.058	0.888	0.291	25.49**	0.251
X_5: Non-irrigated cropped area	2.044	0.425	27.63**	0.312	0.428	0.251	13.84**	0.154
X_6: Fertilizers	4.380	0.361	34.36**	0.360	4.536	0.650	55.37**	0.421
X_7 Farm buildings	7.655	0.410	43.86**	0.418	6.569	0.282	26.81**	0.261

**Significant at 0.01 level.

TABLE 12

REGRESSIONS RELATING LAND-USE AND LABOR INTENSITIES TO ACCESSIBILITY

Dependent Variable	Wheat Region				Rice Region			
	Constant	Slope	F-Ratio	R^2	Constant	Slope	F-Ratio	R^2
a. % land cultivated per farm . . .	96.901	2.330	5.021*	0.076	94.031	1.941	4.084*	0.051
b. ln (labor per hectare of cultivated land per farm . . .	0.522	0.149	6.030*	0.091	1.234	0.154	7.393**	0.089
c. ln (cropped hectares per hectare cultivated per farm . . .	0.569	0.101	21.000**	0.257	0.556	0.020	2.300	0.029

SOURCES: (a) Percentage of farmland under crops was taken from Buck, LUSV, p. 40; (b) ln (labor per hectare of cultivated land) was derived from ibid., pp. 297, 286; (c) ln (cropped area per hectare of cultivated land) was from ibid., pp. 423-24.

*Significant at 0.05 level.

**Significant at 0.01 level.

in an average farm and accessibility. It tells of the tendency for farm operators to expand production at the extensive margin of cultivation (i.e., by augmenting the input of inferior land) as accessibility improved.[1] It also partly explains the positive association between average crop area per farm and accessibility in table 10.[2]

However, production expansion at the extensive margin was attended by increasing marginal cost of land improvement, which made expansion at the intensive margin of cultivation a progressively more economically attractive option. Complementary inputs, especially more mobile factors of production like labor, would be substituted for land as accessibility improved. The result is a rise in the level of land-use and labor intensities. Part (b) of table 12 gives the empirical results in support of the expected rise of labor/cultivated land ratio with increase in accessibility, while the results in part (c) corroborate the positive correlation between land-use intensity, as measured by the ratio of crop-hectare area to cultivated area, and accessibility. The land resource constraint therefore rendered production more elastic with respect to cropped area than cultivated area given similar changes in accessibility conditions.

The heterogeneity in the unit of measurement does not allow meaningful comparison between the estimated factor elasticities in table 11 (except between the land variables). Useful insights,

---

[1] It may be assumed that land qualities were randomly distributed with regard to the accessibility of localities and also among farms in a given locality. As the average percentage of farm area under cultivation increased in an average farm, more and more land of increasingly inferior quality would thus be brought into use (i.e., the average farm had its own Ricardian margin).

[2] Average crop area per farm is the product of average farm size and the average percentage of farm area under cultivation. If farm size were randomly distributed among localities, the positive association between crop area per farm and accessibility would reflect the positive relationship between percent farmland under crops in a farm and accessibility as identified here. A regression was run relating average farm size in the sample localities to accessibility, which found no significant relationship between the two variables in the Wheat Region. In the Rice Region, however, a positive correlation was found between farm size and accessibility with significance confirmed at the 1% level of confidence. A possible explanation for this seemingly anomalous result with the Rice Region is that sampled localities in this region included some extremely isolated agrarian communities (e.g., in southwestern provinces) which were characterized by very low population mobility. High migration costs and the absence of employment opportunities in the undeveloped urban sector of these areas discouraged migration and contributed to rapid multiplication of rural population. For example, rural population growth rates in the interior provinces of Ssuch'uan, Kuanghsi and Yunnan are alleged to be twice as fast as those in provinces of Central and East China between 1900 and early 1930s (see Perkins, Agricultural Development, appendixes A and E). High population density and the tradition of equal inheritance of land among male heirs almost certainly resulted in very small farms in these low-accessibility regions.

however, may be gained when the effect of accessibility change on
input use is expressed in terms of the resulting contribution to
output. A measure of the relative contribution of factor inputs in
this respect is given by the product of the elasticity of output with
respect to a particular input and the elasticity of that input with
respect to accessibility. These elasticity estimates, which may be
called factor-specific output elasticities with respect to access
accessibility, are given in table 13. Each of these elasticity
figures measures the percentage change in output effected by the
change, ceteris paribus, in the employment of the specific factor of
production in response to a one percent change in accessibility.

Given the same degree of change in accessibility, it can be
seen that cropped area had the largest effect on output. It is also
clear that, in conformity with the predominant cropping system in
the agricultural region, dry-farming land had a larger marginal
contribution to output than irrigated land in North China. The
reverse, however, is true in South China. Also expected is the
high marginal contribution of cultivated area to output, due simply
to its inelasticity in supply.

A seeming anomaly in these elasticity estimates is that labor
input differed sharply in its marginal contribution to output in
response to accessibility change in the Wheat and Rice Regions.
For a one percent decline in total transfer cost, the incremental
gain in output as effected by the change in labor use was some 60%
lower in the Rice than in the Wheat Region. The basic reason for
this disparity in labor performance lies in factor proportion
differences in the two agricultural regions. Because of much faster
growth in rural population but slower expansion in cultivated
acreage since the late 19th century, the average man/land ratio in
the Rice Region was about double that in the Wheat Region in the
early 1930s.[1] The consequent low productivity of labor and smallness
of farm holdings in the Rice Region imposed an economic limit on the

---

[1] Between 1893 and 1933, cultivated acreage in the Wheat Region increased by about 3.2% and in the Rice Region by about 2.7%. Rural population, however, expanded more than 33% during the same period in the Rice Region, as compared to Wheat Region's 5% (based on data in Perkins, Agricultural Development, appendixes A & B). As a result, rural population density in the Rice Region was estimated at 2,017 persons per square mile of cultivated land in the early 1930s, more than twice as high as the Wheat Region's 940 persons per square mile (see Buck, Land Utilization, p. 362).

effective use of labor in augmenting production.[1] In effect, low and sharply diminishing returns to labor in farm production had encouraged the more liberal use of complementary factor inputs such as fertilizers and farm implements in the production process.[2] Owing to their largely home-produced nature, these productive inputs may be procured at relatively low costs.

TABLE 13

FACTOR-SPECIFIC OUTPUT ELASTICITIES
WITH RESPECT TO ACCESSIBILITY

Input	Wheat Region	Rice Region
Labor*	0.0939	0.0351
Cultivated area	0.0848	0.1292
Cropped area	0.1257	0.1547
Irrigated cropped area	0.0056	0.1231
Non-irrigated cropped area	0.7690	0.0090
Fertilizers*	0.0697	0.0633
Farm buildings*	0.0432	0.0404

*Calculation on these factors was based on the simple average of the output elasticities derived from the three regressions in table 10.

Allocative Efficiency and Accessibility

Under perfectly competitive market conditions, profit maximization requires that all variable factors of production be employed up to the level at which the value of marginal product (VMP) and factor price are equalized for each factor.[3] This

---

[1] A comparison of the marginal products of labor is given in table 21, column (2), in appendix C. The average size of farms in the Rice Region, according to Buck, was only 3.11 acres, as compared with 5.63 acres in the Wheat Region in the early 1930s (see ibid., p. 272, table 7).

[2] This is reflected in the relatively high elasticity estimates of the variables fertilizers and farm buildings in table 11 (under Rice Region). Since the omitted variable, farm implements, appears to be highly correlated with the farm building variable, it is expected to have high elasticity estimates as the building variable.

[3] Let X_1 and X_2 be two factor inputs in the production function $Q(X_1, X_2)$ with given market prices r_1 and r_2, respectively. Then the profit function is

$$\Pi = pQ(X_1, X_2) - r_1 X_1 - r_2 X_2$$

where p = output price. Maximizing profit with respect to inputs would give

condition is commonly known as allocative or price efficiency in the production process.[1] In reality, however, this condition may not be met due to market imperfections and, particularly in traditional economies, to institutional rigidities.

If only because increasing population pressure would render shifts toward more efficient utilization of resources economically rewarding, the evolution of Chinese agriculture within the relatively static environment of centuries-old farming technology would appear to be an adequate basis for regarding the achievement of a high level of allocative efficiency as a natural consequence of learning by doing. Attenuated by high internal costs of transport and strong resistance from the Chinese, the diffusion of western economic impacts on traditional China since the 1840s was slow and incapable of infusing drastic changes in the environment which could have upset the peasantry's adjusting equilibria. Indeed, the more visible changes in the rural environment were those associated with urbanization, railway and motor road construction, and an expanding array of imported consumer goods--changes that plausibly promoted rationality on the part of the farm operator in his allocation process.

These changes in the economic environment in association with the modernization process also brought up a perhaps more subtle aspect of allocative efficiency in the farm sector; that the peasant's objective conditions of achieving an allocative optimum might actually be improved by increases in accessibility as a result of greater economic incentives to participate in the market economy, improved and more accessible market information, greater factor mobility, and exposure to an enlarged market, even if on a local scale, which led to more competitive market conditions.[2] Spatial variation in the degree of allocative efficiency by individual production units, however, are not in conflict with a high level of efficiency observed in any aggregate national or subnational geographic unit. Obviously

---

$(\partial \Pi / \partial X_1) = 0 \quad p(\partial Q / \partial X_1) = r_1$

$(\partial \Pi / \partial X_2) = 0 \quad p(\partial Q / \partial X_2) = r_2$

i.e., VMP = factor price for X_1 and X_2 respectively.

[1] See Yotopolous and Nugent, pp. 71-86.

[2] Skinner, in a justly famous study, presented much evidence that suggests transport modernization, particularly the building of railways, was a potent force that led to an expansion of marketing areas and intensified trade, but a reduction in the number of standard markets (see G. W. Skinner, "Marketing and Social Structure in Rural China, Part II," Journal of Asian Studies 24 [Feb. 1965]: 195-228).

the mean may encompass individual magnitudes that diverge from the central tendency. Appendix C, which summarizes the estimation procedures used in the following analysis, contains an empirical test of the allocative efficiency of farm labor in the two Crop Regions and finds little evidence to refute the postulate of highly efficient allocation of labor in the agricultural production process.[1] The more interesting question for this study, however, is whether accessibility had also exerted some influences on the level of allocative efficiency in peasant farms by modifying conditions in the objective environments, given the general tendency for all farms to optimally allocate available productive resources?

In order to address this question empirically, two figures, VMP (value of marginal product) per man-workday and the daily cost of labor (wage plus board), were estimated for each of the 127 localities for which data are available. The estimation of VMP per man-workday was based on results of the production function analysis as given in table 10 above. The daily cost of labor was estimated from the wage data collected by Buck. Estimation procedures and sources of relevant data are described in appendix C.

A regression was run relating the logarithm of the absolute difference between VMP per man-workday, an estimate of labor productivity, and daily labor cost, a measure of the opportunity cost of farm labor, to the logarithm of accessibility. The results are shown in table 14. In the Wheat Region, where the distance barrier was much more keenly felt, the expected relationship is highly significant. In the Rice Region, however, the association is not as clearly revealed, though the sign of the estimated coefficient is still as anticipated. These results, when taken as a whole, do tend to support the postulate of improved efficiency in resource allocation in more accessible areas. Thus as transport improvement increased the accessibility of a locality, it afforded strong impetus to increased agricultural production both by encouraging the added employment of productive resources and by changing the objective environment of production so that available resources might be more efficiently utilized.

Railways and the Growth of Agricultural Output

The beneficial impact of railway development of agricultural productivity is the subject of the hypothesis examined in this section. It is patent enough that, when rephrased in terms of

---

[1] Insufficient data prevented the testing of the allocative efficiency of productive resources besides labor.

railways' contribution to accessibility improvement, the present hypothesis (hypothesis II) is no more than a logical extension of hypothesis I which has already been fully explored in the empirical domain in the preceding section. The conclusion reached in the above analyses is that increasing access to market simultaneously conveyed economic incentives and opportunities which encouraged agricultural production through added employment of productive resources as well as their more efficient utilization. In this perspective the gain in agricultural output on account of railway development may be understood by way of the measured change in accessibility ensuing from railway construction.

TABLE 14

REGRESSION RELATING LABOR ALLOCATIVE EFFICIENCY TO ACCESSIBILITY

Region	Constant	Slope	F-Ratio	R^2	Sample Size
Wheat	-5.353	-0.837	27.926**	0.325	60
Rice	-6.362	-1.105	0.610	0.010	67

NOTE: The regression equation is $\ln d = \ln c + h \ln A$, where d = absolute difference between VMP per man-workday and daily cost of farm labor, A = accessibility index, h = slope, or percentage change in d effected by one percent change in A, and c = constant. For derivation of dependent variable d, see appendix C.

**Significant at 0.01 level.

There is, however, an important drawback to this approach. Though it is true that better transport facilitated geographic extension of the internal and external market by lowering transport costs and widening the opportunities for profitable exchange, different transport technologies obviously differed in their capacity to effect such transformations. A railway network, for example, would be capable of forging a national market, whereas the building of a dirt road might affect little but the immediate local economy. Differences in the transport savings to the individual trader-producer, especially on a per trip basis, might be inconsiderable in the two cases, but agricultural producers who enjoyed the new, open access to a large regional or even the national market on account of a railroad would likely respond with a much larger real output than if they only had the blessings of a new dirt road. First there was the scale economy associated with rail transport, which promoted the division of labor and the consequent transfer of resources from the transport sector back into the agricultural sector.[1] Second, the production of industrial crops depended very much on the ease of access to major industrial centers which, in prewar China, were also principal rail terminals. The accessibility to rail-heads hence bore considerable weight in the decision to produce industrial crops on the part of cultivators. Third, internal (intraregional or local) market stability is closely related to the extent of external (extraregional) market integration. As an expanding railway network bound together markets in different regions, they also served to stabilize price fluctuations at regional and local markets.[2] This, as mentioned before, helped to raise the supply elasticity of agricultural output. Last, but not least, is the self-reinforcing effect of productivity and income improvement. If railways conferred benefits on peasants far more than a dirt road or even a dredged stream, the enriched peasants who enjoyed the service of rail transport would have ever higher incentives to increase production and marketing precisely because they had more margins to work with and were willing to take greater risks in reallocating resources away from subsistence production.

An implication of this discussion is that the previously estimated elasticities of farm output with respect to accessibility underestimated the response of farm output to railway-induced accessibility improvement. The more sample observations in the cross-section analyses above which were unaffected by rail transport, the larger would be the resulting downward bias. This problem, however, appears to be relatively minor in the Wheat Region, where more than 75% of all sampled localities actually made use of rail transport in the marketing process (see table 15 below). In the

---

[1] As is characteristic of most traditional economies, the traditional transport sector in prewar China was a huge consumer of productive resources. One estimate put the number of laborers directly engaged in the industry in the late 1920s at about 20 million (besides dependents), or close to one-fifth of the country's entire labor force at the time (see Cheng Ming-ju, p. 34). As long as the marginal productivity of labor and other resources, notably animal labor, remained positive (and there is no evidence to the contrary), agricultural production could always gain from the employment of resources released from the transport sector. The competition for scarce resources by the transport and agricultural sectors was often acute, as can be discerned from this observation: "In Northwest China, Shanhsi and Shenhsi, there is no traffic in June and July as the harvesting calls for the use of animals . . . The goods coming along in December are often six months overdue. Some have been lying in inns or in farmers' houses for weeks on end, while the carter . . . has been sowing, reaping, and sorting his autumn crops" (ibid., p. 39).

[2] A comparative study of grain price movements in the lower Yangtze (delta) area found that seasonal prices of wheat and rice were much more stable in the early twentieth century than they were 200 years ago. Chuan and Kraus suggested that the difference in pattern may be attributed to the availability of rail transport which gave southern Chiangsu rapid access to northern grain in a way that was simply impossible two centires earlier (see Chuan and Kraus, pp. 22-23).

Rice Region, where no more than 20% of the sampled localities were affected by railways as far as long-distance marketing is concerned, the bias might be considerable.[1] The problem in the Rice Region is further aggravated by the inclusion of relatively isolated provinces in the Southwest into the analysis. The lack of exchange relationships between markets in these interior territories and those in the more commercialized East in fact acted much to discourage the production of higher-valued industrial crops in the interior regions.

Bearing in mind that the measured response of farm output to rail-induced accessibility change represents a lower-bound estimate, we now ask the question to what extent had railway construction contributed to accessibility improvement and regional agricultural output? A simple approach to answering this question is to compare the mean accessibility index of those localities that were affected by railways with the mean accessibility index of these same localities in a no-railway situation. The validity of this approach rests on the counterfactual proposition that the localities, if deprived of railway service, would continue to market their products to the nearest regional markets. The nearest regional markets in the no-railway situation, however, need not be the same ones in the actual situation with railway service.

Because the economically feasible size of market areas of many localities would be smaller, and in some cases dissipated entirely, in the no-railway situation, the counterfactual assumption of continued marketing without railway service disguises the effect of railways on market expansion and adds to the aforementioned bias. Also, the actual reduction in transport cost would likely exceed the computed change as rail competition might have forced downward adjustments in the transport rates of alternative, traditional modes of transport. These considerations render the following results lower-bound estimates of the effect of rail-induced accessibility change on farm output.

The estimation procedure is summarized in table 15. In the Wheat Region, a 64.5% increase in average accessibility was observed

---

[1] It may be noted that most of the communities along major railway lines in the lower Yangtze delta region actually found rail transport rates somewhat, though not substantially, higher than traditional transport using river and canal junks. Since the accessibility index was computed with respect to transport rates, these communities had high index values but were considered not making use of railways in their marketing processes. In fact, much evidence points to the importance of railways in agricultural marketing, especially in regard to commercial crops (see chapter III). The bias therefore stems in part from the omission of indirect benefits (e.g., time and risk savings) in the calculation of the accessibility index.

TABLE 15

ESTIMATED FARM OUTPUT RESPONSE TO RAIL-INDUCED
ACCESSIBILITY CHANGES

	Wheat Region	Rice Region
a. The before-rail mean accessibility index of affected areas	0.020	0.070
b. The after-rail mean accessibility index of affected areas	0.033	0.092
c. Percentage change in mean accessibility index	64.5%	31.2%
d. Sample size of affected areas	48	19
e. Sample size of affected areas as a percentage of total sample size	76.2%	19.2%
f. Percentage increase in affected area output due to accessibility change	17.1%	7.8%
g. Total pre-rail output of affected areas as a percentage of regional pre-rail output	76.3%	22.1%
h. Percentage increase in regional output ([f] x [g])	13.0%	1.7%

for 48 localities which were affected by railway construction (row [c]).[1] The corresponding figure for the Rice Region with 19 localities was 31.2%. Substituting the with- and without-railway mean accessibility indexes into eq. (1) (table 9), it was found that the average increases in the farm output of affected localities were 17.1% and 7.8% in the Wheat and Rice Regions respectively (row [f]). In other words, for the sampled localities which made use of rail transport in agricultural marketing, the construction of railways was able to boost average farm output by at least 17% in the Wheat Region and 8% in the Rice Region. It is evident from this result that, even abstracting from the difference in the size of the regional network, railways were a far more important agent of agricultural change in North than in South China.[2]

---

[1] The calculation of the before-rail accessibility indexes follows closely that of the after-rail indexes (as outlined in appendix B), except that now a railway route would be replaced by the next cheapest non-railway route that leads to the nearest regional market.

[2] Obviously the main reason for the large difference in railways' regional contribution is the competitiveness of the water transport system in South China. Market accessibility is not only affected by the transfer rate pertaining to alternative modes of transport, it is also related directly to the distance of the haul. Since South China had a much denser settlement pattern than North China, the average haul to the regional market was shorter in South China. The existence of low-cost water transport and the short-distance usually involved in agricultural marketing greatly dampened the impact of railways on farm production in the Rice Region.

The regional disparity in the effect of rail transport is even more glaring when regional, instead of affected-area output, is considered. Here the size and density of the railway system become prime considerations, and the deficiency of the prewar railway system in China, particularly in South China, is hardly concealed. As shown in row (h) of table 15, the estimated increase in regional farm output amounted to 13% in the Wheat Region, but it was not more than 2% in the Rice Region. Needless to say, had the railways penetrated into such interior regions as Kuanghsi, Kueichou, Ssuch'uan and Yunnan, and into transportationally deficient areas such as Fuchien and southwestern Hunan, the effect of railways on farm output in the Rice Region would have been more conspicuous.

Until now the attempt to relate gains in farm output to railway development has relied on inference. The merit of this inferential approach lies in the elucidation of the economic mechanisms which underlay the response of the prewar Chinese agrarian economy to expansion of trade opportunities attendant upon railway development. Whether communities that capitalized on the economic opportunities brought about by railways had in fact gained in output is an unanswered question, however. No direct examination of this question is indeed possible because of the absence of cross-section time-series data. Nonetheless the remainder of this chapter is devoted to an exmination of available evidence which may provide partial answers to the question.

In what follows a comparative analysis is carried out on the relative levels of farm output and land productivity in rail-adjoining and non-rail-adjoining localities. This is done under the assumption that no ex ante difference in output and productivity levels existed in the two groups of communities. The next section examines the effect of railways on actual changes in farm productivity in the light of time-series data on farmland values. Results from both analyses fit neatly into the conclusion of the previous section: there was a definite relationship between railway development and improvement in agricultural output.

The comparative analysis of rail-adjoining and non-rail-adjoining localities (hereafter railway and non-railway localities, respectively, for short) begins with a sampling procedure that statistically control the difference in the average accessibility as well as ex ante (or pre-railway) production levels of the two samples. By choosing samples according to a common range of dispersion with respect to the accessibility index, accessibility-induced difference in output levels can be minimized.

However, the sizes of the samples thus chosen are small enough to render further adjustments for disparities in pre-railway conditions impossible. Fortunately this latter adjustment is found to be unnecessary. This is because existing evidence does not support the hypothesis of uniform directional difference in output and productivity levels of farms in the two groups of communities in the pre-railway situation--presumably a general supremacy of the railway localities over the non-railway localities.

Because of the absence of systematic efforts to record key variables such as farm output and income in Imperial China, the hypothesis of significant ex ante difference in the output of railway and non-railway areas cannot be tested directly. Instead, indirect indicators of farm output and income, such as population density and land-tax rates, have been used for the purpose. In appendix D, population density and land-tax rates are derived from data contained in various historical sources and are employed in an examination of the possible extent to which the hypothesis of no significant ex ante difference in the productivity of railway and non-railway areas is acceptable.

Specifically, population density statistics by hsien were compiled from the gazetteers of Hopei, Shantung, and Chechiang for intra-provincial comparisons.[1] These three provinces were selected both because the historical data are readily available and because these provinces had relatively large railway mileage by the early 1930s. Although the dates of compilation of the gazetteers were not the same, they contain historical statistics which allow data pertaining to a similar time period, from the 1870s to early 1880s in this case, to be generally available. More important, however, is that the density comparisons were made for railway and non-railway localities (defined according to early 1930 status) within the same province, thus providing control on the non-comparability between provincial records.

Unfortunately, the population registration system in late Ch'ing was beset with much inconsistency and confusion.[2] The problem was particularly severe in the middle and lower Yangtze provinces (including Chechiang), which suffered great population loss during

---

[1]The local and provincial gazetteers of China are the most important source of historical information, both quantitative and qualitative, on the local areas of China. These gazetteers were compiled and updated by local authorities at irregular time intervals and represent official records of the history, economy, culture, geography, and politics of the local area.

[2]See Ping-ti Ho, Studies on the Population of China (Cambridge, Mass.: Harvard University Press, 1959), pp. 65-73.

the Taiping and Nien wars in the 1850s and 1860s. Even in areas
less affected by wars the energy of the local officials was consumed
in the more urgent matters of raising revenue at a time when the
country was pinched with economic woes.[1] As a result population
statistics in late Ch'ing were subject to uncertain margins of error.
Even though the statistical test on population density carried out
in appendix D (table 22) indicated no significant difference between
railway and non-railway area densities in both Hopei and Chechiang
(Shantung passed the test marginally), the results must be taken
with reservation.

A more reliable test outlined in appendix D rests on local
land-tax statistics, which were usually well-kept by local and
provincial officials and, others things being equal, would provide
some indication of the relative productivity of local farmlands. As
discussed more fully in appendix D, changes in the rate of land
taxation in late Ch'ing were not commensurate with progress in
productivity, but tended to lag in time. Even so, such rate changes
were seldom incorporated as part of the main tax but rather appeared
as surtaxes under a variety of titles. In an important study,[2] Wang
Yeh-chien was able to estimate the real rate of commutation in
some 20 provinces. Though Wang's primary source of data was a
collection of provincial financial reports prepared in the mid-1900s,
the possibility of time lag in land-tax adjustments renders such
statistics useful in depicting pre-1900 conditions, when most of
China's railways were not yet built.

Basing on data from Wang and other local gazetteers, mean
difference tests were applied to local land-tax rates in six
different provinces: Shantung, Hopei, Chechiang, Honan, Shanhsi,
and Anhuei, which all displayed relatively large railway mileage by
the early 1930s. A similar test was also applied to the sample of
railway and non-railway localities which will be used in the
following analysis. In both cases the hypothesis of no _ex ante_
difference was accepted. In other words, land-tax evidence supported
the assertion that railway routing was in general not dictated by
the pattern of local farm productivity. Although this result must
be interpreted in light of the uncertain extent to which land
taxation rates, even with surtaxes taken into account, corresponded
to the actual level of land productivity, the evidence is in agreement

---

[1] Ibid., p. 71.

[2] Wang Yeh-chien, An Estimate of the Land-Tax Collection in China, 1753 and 1908 (Cambridge, Mass.: Harvard University Press, 1973).

with the conclusion drawn from a direct examination of actual productivity levels for a number of major commercial crops, which will be discussed in chapter V.

Provided that the hypothesis of no ex ante difference is accepted, a difference-of-means test would be able to tell whether railway development had indeed appreciably raised the average output level in railway areas. Note that the failure for such a difference to occur does not imply that the effect of railways on local farm production was nil.[1] However, if this difference is large enough to be detected statistically, it will testify to the strong effect of railways on raising farm output.

The sample of railway localities was chosen according to the criterion that the average marketing distance between the county seat (hsien-ch'eng) and the nearest rail-head should be no more than the average local marketing distance, as reported in Buck,[2] using water or the most common mode of land transport. Non-railway localities were confined to those with an accessibility index falling within 99% of the normal interval spanned by the railway sample.[3] The size of the railway sample was 27 and 15 in the Wheat and Rice Regions respectively. The sample size of the non-railway sample was 23 in the Wheat Region, and 48 in the Rice Region. A map of the geographical distribution of the sample localities appears in figure 9.

To test the hypothesis that farm output was significantly higher in the railway samples, two analysis of variance (ANOVA) tests were undertaken. The dichotomized nominal variable in the tests was

---

[1] First, smallness of sample would amplify the effect of intrasample variance so that only a strong difference in sample means would yield statistically significant results. Second, there is the possibility that pre-railway productivity was actually higher in some non-railway areas (see discussion in appendix D). Unless the effect of railways on farm productivity was uniformly strong across the sample, the resulting difference between sample means would be insignificant.

[2] See LUSV, p. 344. If a locality had access to river transport from the hsien capital to the rail-head yet no water distance was reported, the average water distance would be determined in proportion to the ratio between the transfer rates of junks and the most common mode of land transport. Some excessively long distances reported by Buck in local marketing by boats in the Rice Region were also substituted by average distances calculated in this fashion. The most common mode of land transport was determined by the frequency different modes were reported by Buck in an agricultural region (or province if appropriate).

[3] That is, the feasible range of accessibility of the non-railway sample would be $\bar{x} + 2.58\sigma$ (assuming normal population distributions), where $\bar{x}$ and σ denote the mean and standard deviation of the railway sample, respectively. Taking a lower boundary limit of zero, the computed ranges of transfer costs for the non-railway localities were 0-42.3 and 0-19.5 in the Wheat and Rice Regions respectively.

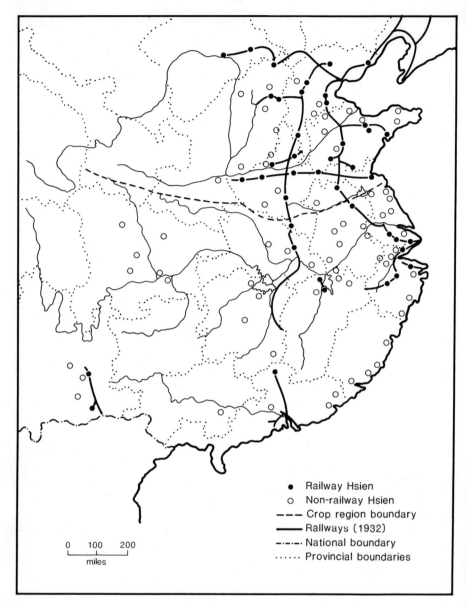

Fig. 9. Location of railway and non-railway hsien

railway versus non-railway localities. Observations were measured
against two independent interval-scaled variables: average farm
output and output per hectare of farmland under crops (yield). The
yield variable, which essentially measures farmland productivity (in
value terms), is an indicator of the average value of agricultural
land--the subject of investigation in the following section.

The results of the ANOVA tests are summarized in table 16.
Except for the yield variable in the Wheat Region, the railway
localities consistently displayed higher mean values which are
significantly different from those of the non-railway areas. The
average farm output of the railway sample in the Wheat Region was
28% higher than that of the non-railway sample. In the Rice Region,
the difference was about 27%. The average yield in railway areas of
the Rice Region was 13% higher than in the non-railway counterparts.
In spite of the statistically insignificant difference, the railway
loclaities of the Wheat Region also showed an average yield that was
4% higher than that in the non-railway localities, a result in general
consistency with the findings above.

TABLE 16

ANOVA TESTS OF RAILWAY EFFECTS ON
FARM OUTPUT AND LAND YIELD

Region	Average Output per Farm			Yield		
	Mean Rail	Mean Non-Rail	Computed F^a ($n_1=1$, $n_2=d$)	Mean Rail	Mean Non-Rail	Computed F^a ($n_1=1$, $n_2=d$)
Wheat	2,625	2,048	8.49(d=49)**	1,179	1,131	3.25(d=49)
Rice	4,433	3,498	4.88(d=62)*	3,822	3,389	4.21(d=62)*

^{a}n_1 and n_2 denote between-group and within-group degrees of freedom, respectively.

*Significant at 0.05 level.

**Significant at 0.01 level.

The Effect of Railways on Land Values

Railway construction raised land rents, and thus land values,[1]
of farmlands because of the upsurge in the marginal productivity of

---

[1]The relation between land rent and land value (or land price) can be expressed in terms of the familiar capital value formula:
$$P = \sum_{t=1}^{t=n} [r/(1+i)^t] \qquad (t = 1, \ldots, n)$$

land as transport conditions improved. The immediate impact emanated from increasing farm prices at a pace commensurate with savings in transfer cost. Improvement in farmgate prices raised the value of the marginal products (VMPs) of production factors. Consequent expansion in the use of complementary inputs in the production process to realize higher levels of output would further boost the value of cultivated land, provided that productive land resources were of inelastic supply.

It is unlikely, however, that transport-induced rises in agricultural land value are a one-shot phenomenon. This is particularly true in traditional economies where alternative industrial employment opportunities remained scarce. In the flush of a railway-influenced prosperity, rural communities bordering on axes of railway construction may experience rapid increases in population density as in-migration intensified. Although improved transport conditions invariably reduced rural-urban migration costs, sluggish growth in industrial employment in urbanized areas would likely result in an equilibrium level of rural population density along railway routes that was perceptibly higher than before.[1] The result was upward pressure placed by the increased population on the value of cultivated land.[2]

The effect of railway construction on agricultural land values may therefore encompass a short-term and a longer-term component.

---

where P, r and i denote the present value of land, annual rental income (assumed to be constant), and rate of interest respectively. The time index $t = 1, \ldots, n$ may be taken to indicate the expected time horizon within which the annual income stream would flow.

[1] For a discussion of the empirical relation between population density, the availability of alternative urban-industrial employment opportunities, and the value of agricultural land, see Colin Clark, The Value of Agricultural Land (Oxford: Pergamon, 1973), pp. 15-19.

[2] Given the Cobb-Douglas production function with constant returns to scale

$$X = L^{\alpha} N^{1-\alpha} \qquad (\alpha < 1)$$

where X, L and N denote output, land and labor respectively, and α denotes the constant share of land in output. The marginal product of land (land rent), r is:

$$r = \partial X/\partial L = \alpha L^{\alpha-1} N^{1-\alpha}$$

Denoting proportional changes with respect to time by an asterisk (*), we have

$$(1/r)(dr/dt) = r^* = (\alpha-1)(L^* - N^*)$$

Thus if labor growth is faster than land expansion, i.e., $N^* > L^*$, then $r^* > 0$ (since $\alpha < 1$); the magnitude of r^* being in proportion to the change of $(L^* - N^*)$.

In the immediate years following the construction of the railway, farmland values would be anticipated to jump discretely upward from pre-railway levels. Over a longer time span, however, the trend of increase in land value would be maintained at a higher level than in the pre-railway period, and that communities adjacent to the railway would experience a faster growth in cultivated land value than in more distant areas, as long as drastic increases in alternative industrial employment opportunities did not occur.

Using the land price index time-series data gathered by Buck,[1] these propositions can be examined in the empirical context of prewar China. For the test of discrete jumps in land values which followed railway construction, a sample of 19 railway hsien and 24 non-railway hsien has been selected.[2] The percentage difference between three-year average land values (indexed) immediately before and after the first year of full operation of an adjacent railway was computed for each locality in the sample. From the results in column (1) of table 17, it can be seen that railway localities as a group did exhibit more drastic changes in farmland values in response to the construction of railways than did non-railway localities. The difference was significant at the 1% level of confidence.

For the test of trend slope difference, 20 localities constituted the railway sample and 58 were in the non-railway sample. Two trend growth rates were computed for all localities in the railway sample: one pertaining to land value movements before the date of full operation of the adjacent railway; the other pertaining to movements after the date of railway operation. The group-average rate of growth in land values was found to have increased by about 2% per annum from a before-rail average rate of 3.5% to an after-rail average rate of 5.6% per annum. This provided evidence in support of the hypothesis that land values in railway-affected areas increased their rate of growth upon the completion of railway lines.

The after-railway trend growth rates in the railway areas were further subjected to an ANOVA test against the average trend growth rates in the non-railway areas. The result of this test is given in column (2) of table 17. With a statistical significance established at the 5% level of confidence, the result supported the postulate that value of agricultural land should have increased faster in railway-adjacent areas following the construction of the railway than in areas not much affected by such transport changes.

---

[1] LUSV, pp. 168-69.

[2] Because of the small size of the sample, tests were conducted for pooled (Wheat and Rice Regions combined) samples only.

TABLE 17

ANOVA TESTS OF LAND-VALUE MOVEMENTS IN
RAIL AND NON-RAIL LOCALITIES

	Percentage Change in* Land Values Before and After Rails (1)	Land Value Trend** Growth Rates (2)
Railway Localities	19.4% (n=14)[a]	0.056 (n=20)[b]
Non-Railway Localities	11.5% (n=28)[a]	0.034 (n=58)[b]
Computed F	7.53	5.63
Significance Level	0.01	0.05

*Percentage change in farmland values (indexed) between three-year averages immediately before and after the first year of full operation of an adjacent railway line.

**Estimated land-value trend growth rates by time-series regressions of the form $\ln V = a + \gamma\, t$ where V = indexed land value, γ = trend growth rate estimate, t = time index, and a = a constant parameter. Railway locality series began from the date of full operation of an adjacent railway line to the year 1926, the base year of the indexed series. Non-rail locality series were inclusive of all data entries and terminated again at 1926. In both cases only series with no less than fifteen observations were chosen for analysis.

[a] Sample sizes: 8 and 6 localities in the Wheat and Rice Region, respectively, in the Railway sample, and 14 localities each in the two groups in the Non-Railway sample.

[b] Sample sizes: 12 and 8 localities in the Wheat and Rice Region, respectively, in the Railway sample, and 24 and 34 localities in the Wheat and Rice Region, respectively, in the Non-Railway sample.

Summary

In this chapter empirical evidence has been amassed and analyzed to examine the hypothesis that railway development in prewar agrarian China stimulated growth in agricultural output. The theoretical underpinning of an expected causal relationship between rail transport development and agricultural growth rests on the often observed responsiveness of farm operators, in low and high-income countries alike, to profit incentives and new opportunities for trade. As farm prices rose and markets expanded with improving transport conditions on account of railway development, peasants responded with added employment of productive resources and their better, or more efficient, utilization. The result was significant gains in real farm incomes.

The economic mechanism of the rational response of farmers in prewar China to railway-induced improvement in transport conditions has been examined in terms of a postulated positive relationship between market accessibility and expected farm output. Empirical estimation of the relationship has drawn on cross-section data gathered by J. L. Buck during the early 1930s. By showing that accessibility improvement benefited farm production and contributed to economic efficiency in the agricultural production process, and that railway development increased the market accessibility of rural communities, the analysis was able to confirm the positive role of railways in China's agricultural growth.

The exact extent to which railway development affected regional agricultural output, however, was underestimated by the above approach. The computed rate of decline in transport cost due to railway construction neglected downward adjustments that probably occurred on the transfer rates of traditional competitive modes of transport. It also slighted the potential effects of railways on the expansion of extraregional markets and the accessibility to large industrial centers. These effects tended to confer special economic advantage on rural areas which were close to railway lines.

In order to find out whether these economic advantages were actually exploited and translated into higher farm outputs, a comparative test of farm output and land productivity was carried out between railway and non-railway localities. A separate test, described in appendix D, established the precondition that no ex ante (before-rail) difference in the agricultural productivity level of the railway and non-railway areas prevailed. Given this result, the comparative productivity test lent support to the observation that rail-adjacent areas generally exhibited higher output and land yield (by value) per farm than non-rail-adjacent areas.

A final test of the differential performance of agriculture in railway and non-railway areas made use of the cross-section time-series data on land values collected by Buck. A closer analysis of these data revealed that railway construction not only entailed short-term surges in local land values, it also provided a strong drive toward more rapid growth in land price over the long-run as in-migration of population placed upward pressure on land productivity. The axes of railway construction, in other words, represented the axes of change in prewar agrarian China.

CHAPTER V

RAILWAY DEVELOPMENT AND AGRICULTURAL COMMERCIALIZATION

Introduction

In the opening decades of the present century, the traditional, agrarian economy of China underwent an unprecedented process of agricultural commercialization which underscroed the contemporaneous gain in agricultural output.[1] Cash crops, mainly non-food commercial produce, took up an increasingly larger share of China's cultivated acreage at the expense of subsistence food staples, which also gave way slowly to high-yielding, non-staple varieties such as corn and sweet potatoes as scarce land resources were released for cash crop production.[2] As a result, real output of non-grain crops as a whole expanded by more than 40% between the mid-1910s and the mid-1930s, and its share in total agricultural production rose from 17% to 21%.[3] In contrast, total grain production in real terms increased no more than 8% during the same period, and its share in total output actually declined.[4] Animal husbandry, also a market-oriented farm production activity, had gained in importance as a source of cash income for peasants.[5] Total animal output

---

[1] One estimate put China's total agricultural output in the mid-1930s at about C$15.6 billion, or 15% above the mid-1910 output measured at 1933 prices. See Dwight Perkins, Agricultural Development, p. 289.

[2] China's cultivated area devoted to barley, indigo, kaoliang, millet and rice gradually declined after 1900, whereas that occupied by sweet potatoes, corn, soybean, peanut, rapeseed, cotton and sesame increased (see Buck, Land Utilization, p. 217). In North China, the cultivation of cotton and sweet potato steadily increased, and in Central China peasants used more land for oil-bearing seed crops (see Myers, "Commercialization of Agriculture," p. 183; see also idem, The Chinese Peasant Economy, pp. 190-93).

[3] Perkins, Agricultural Development, p. 289.

[4] Grain output amounted to over 75% of total agricultural production in 1914-18, but it dropped to only about 70% of output in 1931-37 (in 1933 prices) (see ibid.). It may be noted that besides corn and potato, wheat was one of the few grain crops that had above average growth rates. This was because wheat was both a commercial and subsistence crop, and also because wheat can be planted in rotation with cotton.

[5] See Buck, Land Utilization, p. 298.

probably increased by more than 23% in real terms over the two decades prior to the mid-1930s.[1]

To what extent was this quickened tempo of market production in the peasant economy of prewar China attributable to railway development and associated transportation changes? It is the purpose of this chapter to shed light on this question by an examination of the empirical evidence. The results of the previous two chapters, which highlighted the role of rail transport in the expansion of long-distance agricultural commerce and the growth of agricultural output in prewar China, in fact carried strong implication for the role of railways in the process of agricultural commercialization. Nevertheless two considerations render a direct and deeper inquiry worthwhile.

First, the extent to which railways contributed to the process of commercialization in the agrarian economy is an empirical question and must be examined in the light of empirical evidence. Second, agricultural commercialization signifies the enhanced participation of the peasantry in the monetized exchange economy, a process which encouraged the division of labor and the specialization of production. The immediate result was increases in farm income. In the long run, however, agricultural commercialization also occasioned capital formation and subsequent growth in the farm, as well as the non-farm, sector through a variety of linkage effects.[2] Given its

---

[1] Perkins, Agricultural Development, p. 289.

[2] Backward linkages would lead to new investments in input-supplying facilities as income from "exchanged production" is spent on farm capital inputs. Investments thus generated are, however, often limited in the case of traditional agriculture. But such new incomes, when spent on imports, can eventually lead to sustained expansion of import-substituting industries through what Hirschman called "consumption linkages" (Albert O. Hirschman, "A Generalized Linkage Approach to Development," In Essays on Economic Development and Cultural Change in Honor of Bert F. Hoselitz, ed. Manning Nash [Chicago: University of Chicago Press, 1977], pp. 67-98). Such linkage effects were probably instrumental in the early development of China's cotton textile and tobacco industries. Expansion in the households is another linkage which would lead to human capital formation in the farm (i.e., through improvement in health and knowledge, enhanced receptivity toward modern ideas, methods and production, etc.).

Forward linkages lead to investments in output-using facilities, as the growth of marketed agricultural output sustains urbanization and industrial development in the modern sector. The increased demand for transport facilities also encourages further expansion in the transport sector and related infrastructural developments. These in turn have second-round effects on both the agricultural and the industrial sectors as they gain from scale economies of production and exchange, internal market integration, and improvement in marketing efficiency consequent on transport growth. Hirschman further identifies the plausibility of increased tax revenue from improved rural incomes thus produced as "fiscal linkages," which can be an important source of investment capital for the public sector in low-income, predominantly agricultural economies (ibid.).

implication for the economic growth of early modern China, the process of agricultural commercialization warrants a closer study.

A natural point of departure for the present inquiry is the choice of a commercialization index. This is a measure which appropriately gauges the extent to which farm productive efforts are directed at earning a money income.[1] One indicator often used is the marketed surplus, or the proportion of farm output actually marketed. However, marketed surplus coincides with the planned level of cash production only if there are no forced sales immediately following harvest for loan repayment and no speculative retention of marketable output on the part of the farmer. Neither case is likely to hold in reality. Moreover, data on actual cash receipts from marketed farm output in prewar China are hard to come by, and value of output at the farm level for any particular year is impossible to estimate.

For the above reasons the present inquiry will rest on the concept of "marketable surplus," which is defined as total farm output in excess of actual consumption demand of the farm household. Marketable surplus refers to the potential surplus in farm output. It therefore allows changes in the propensity of farmers to participate in the market to be reflected in such circumstantial evidence as expansion in the acreage of cash crops and intensification in the regionalization and specialization of agricultural production. The following section is devoted to an examination of this evidence and their relation to railway development. A subsequent section then analyzes the direct connection between farm marketable surplus and accessibility. The result is interpreted in the light of railways' impact on accessibility, a subject already taken up in chapter IV above.

Regional Specialization and the Expansion of Cash Cropping: Evidence on Railway Effects

There is much evidence that, in agrarian China, expected profitability was a basic determinant of resource allocation toward monetary (non-subsistence) production. Accessibility to markets, which affected the relative prices of marketable output, also had considerable influence on the degree of market orientation of rural producers. A study by Fei and Chang which bears on this issue finds

---

[1] Conceptually, agricultural commercialization involves the development of a two-way tie to the market economy--output and input. The input tie, which signifies the increased purchase of manufactured output and services from the modern sector by the rural sector, may be expected to develop as the output tie strengthens. The lack of empirical data prohibits a close examination of the input-tie in the present study.

market accessibility of considerable importance in the production choice of three villages in Yunnan province.[1] The influence of a nearby market on rural production was also analyzed by Yang.[2] Rawski, in a study of certain South China rural communities during Ming and Ch'ing dynasties (between 16th and 18th centuries A.D.), describes how rural commercialization was intimately entwined with market forces, particularly movements in relative prices, and market accessibility.[3] The marked responsiveness of the Chinese peasant to profit incentives can indeed be found in numerous historical documents. A selection of these appears in a collection of historical materials on Chinese agriculture since the 1840s.[4]

The evidence on the positive price responsiveness of Chinese peasants suggests that the marked expansion of cash cropping in prewar China can be attributed to an improving terms of trade faced by rural producers. Data collected by Buck showed that the farm price index of marketed farm products increased over 200%, or more

---

[1] Hsiao-tung Fei and Chih-i Chang, Earthbound China (Chicago: University of Chicago Press, 1945).

[2] C. K. Yang, A Chinese Village in Early Communist Transition (Cambridge, Mass.: Technology Press, 1959). Similar observations were also made by M. C. Yang in A Chinese Village - Taitou, Shantung Province (New York: Columbia University Press, 1945), pp. 199-200.

[3] Evelyn S. Rawski, Agricultural Change and the Peasant Economy of South China (Cambridge, Mass.: Harvard University Press, 1972).

[4] See Li Wen-chih and Chang You-i, eds., Chung-kuo chin-tai nung-yeh shih tz'u-liao (abbreviated as CKNY). Explicit accounts of price responsiveness of Chinese peasants can be found in the following example excerpts: 1:396 (Customs Report); 1:419 (Chu Tsu-yung, T'ung-shu chung-mien shih-lu); 1:422 (Wang Fung-lou, Chiu-chiang fu-chih); 2:133 (Chien-yeh yüeh-pao); 2:134 (Chang-li hsien chih ching-chi chuang-k'uang); 2:147 (Wah-shang sha-ch'ang mien-ch'an t'iao-ch'a pao-kao); 2:150 (Yang Ta-huang and Chai Yung-chung, Yü-ho mien-ch'ü ti mien-tso); 2:151 (Lo Kuan-ying, Chiang-su wu-hsi hsien erh-shih nien-lai chih ssu-yeh kuan); 2:190 (Chiao Nung-wah, Wo-kuo ts'an-ssu chih hui-ku yu ch'ien chan); 2:193 (Wang Tse-pu, tr., Nan-chung-kuo ssu-yeh t'iao-ch'a pao-kao-shu); 2:198 (Ministry of Industry, Chung-kuo shih-yeh chih); 2:199 (Hupei-sheng chih yen-t'sao); 2:202 (Chinese Economic Journal); 2:203 (Kuang-tung ta-hsüeh nung-hsüeh-yuan, ed., Kuang-tung nung-yeh t'iao-ch'a pao-kao-shu: Ch'ing-yuan hsien); 2:207 (Chi Pin, Nung-ts'un p'o-ch'an sheng-chung chi-nan i'ke fan yung ti ts'un-chuang); 2:210 (Customs Decennial Report, 1922-31); 2:211 (Wah-shang sha-ch'ang mien-ch'an t'iao-ch'a pao-kao; and Wu Ching-wang, Hu-hai-tao ch'ü shih-yeh shih-cha pao-kao); 2:212 (Yang Cheng-chien, Pieh-szu-sheng tsai-ch'ü shih-ch'a chi); 2:213 (Chinese Economic Journal); w:214 (Chi Pin, Nung-ts'un p'o-ch'an sheng-chung chi-nan i'ke fan-yung ti ts'un-chuang); 2:223-24 (Yung An, Ke-ti nung-min chuang-k'uang t'iao-ch'a); 2:425 (Ying-heng chou-pao); 2:426 (Chiang-su shih-yeh tsa-chih; Che-chiang chi ch'a-yeh; and Buck, Wu-hu pai-ling-erh ke t'ien-chia chih ching-chi chi she-hui t'iao-ch'a); 2:427 (Nung-shang kung-pao; Hupei-sheng chih yen-tso; and Wang Tse-pu, tr., Nan-chung-kuo ssu-yeh t'iao-ch'a pao-kao-shu); and 2:428 (Customs Decennial Report, 1912-21).

than 4% per annum, between 1900 and 1930.[1] In contrast, rural import prices, i.e., prices paid by farmers mainly for non-locally produced goods, increased only 117% between 1907 and 1930, or at an average annual rate of less than 3.5%.[2] Buck ascribed the rise in farm prices to the devaluation of silver, the currency in which farm prices were expressed, but he also gave emphasis on the influence of transport improvement, particularly railways.[3] Reduction in transfer costs, in fact, might have accounted for 20% or more of the observed rise in farm prices in North China between 1900 and 1930, though in South China the contribution of transport improvement to farm price increase was probably much less. Appendix E gives a detailed description of the method used to derive these estimates.

The development of rail transport afforded strong incentives for market-oriented farm production because it lowered the cost of long-distance transportation, both in terms of money and time, and thereby improved the opportunity for profitable trade. Increase in trade opportunity made possible reduced reliance on self-produced supplies and allowed the reallocation of resources to commercial, rather than subsistence, production. Often, the increased demand for manufactured consumer goods as farm income improved reinforced the tendency for market production.[4]

Apparently the impluse for trade and increased market production received especially strong stimulus when, by design, the railways established direct communication links to commercial centers which were the dynamic poles of national economic growth. As a consequence of increasing export demand, growth in urban population, and industrial development, agricultural prices at China's terminal urban markets climbed steadily since the 1890s.[5] In many cases, these

---

[1] Calculations based on Buck's indexes as quoted and revised by Perkins, Agricultural Development, p. 363. Buck's index of prices received by farmers for marketed farm products was an average of indexes in some 30 localities in 15 provinces. The index registered a 229% rise between 1900 and 1930.

[2] See ibid.

[3] See Buck, Land Utilization, p. 313. See also idem, "Price Changes in China," Journal of the American Statistical Association (June 1925), p. 241; and idem, Hopei yen-shan-hsien i-pai-wu-shih nung-chia chi ching-chi chi she-hui t'iao-ch'a (An economic and social survey of one hundred and fifty farm families in Yen-shan Hsien, Hopei Province)(Nanking: Nanking University, Department of Agricultural Economics, n.d.), pp. 162-63.

[4] For development of a theoretical model which incorporates this rural-urban interaction in elucidating rural growth and reference to some of the empirical evidence in this regard see chapter VI below.

[5] The wholesale prices of major agricultural exports at China's principal urban markets, notably Tientsin and Shanghai, doubled or even tripled between

steady rises in urban agricultural prices were effectively translated into dramatic jumps in rural farm prices as railways threw open the door to expanded rural-urban trade. In Ching Hsien of Hopei province, for example, the average increase in the prices of ten major farm products jumped from 53% in 1900-1910 to over 170% in 1910-1920.[1] An obvious contributor to this event was the opening of the Tientsin-P'uk'ou railway to through traffic in 1911.

Whether in absolute or percentage terms, the growth of farm prices in Ching Hsien following the opening of the Tientsin-P'uk'ou line exceeded what happened in nearby terminal markets such as Tientsin and other rural areas less close to the railway line.[2] Evidently rail transport had its unique role to play in the experience of Ching Hsien. An even more noteworthy example was that of Changli Hsien in northeastern Hopei province. Local prices of fresh fruits, traditionally an important item of export from the local communities, reportedly increased one hundredfold soon after the completion of the Peking-Mukden railway in the early 1900s.[3] The drastic increase in fresh fruit prices was quickly translated into increases in fresh fruit output and expansion of orchard acreage, including the conversion of low-lying fertile farmlands long devoted to grain crops into tree crops.[4] Similar observations were also reported for the case of peanuts in a rural survey that covered several peanut-growing hsien in six different provinces. According to the survey, in 1900, peanut fields occupied a mere 0.1% of total cultivated acreage in

---

1900 and 1930. The export price of rice, for instance, more than tripled during the period, while that of soybean nearly doubled (see reference given in appendix E).

[1] Based on data in CKNY, 1:562-63. Original prices in copper currency were translated into silver currency (C$) by using currency exchange rates given in L. L. Chang, "Farm Prices in Wuchin, Chiangsu," Chinese Economic Journal 10 (June 1932):498 (see also Buck, "Price Changes in China," p. 240, table II).

[2] According to Tang, cotton price at major markets increased less than 8-16% during 1910-20, or in absolute terms less than C$0.06-C$0.10 per kg. (the lower figure pertains to 1920 as the terminal year and the higher figure to 1919-20 average as the terminal year price)(see Chi-yu Tang, An Economic Study of Chinese Agriculture [n.p., 1924], p. 380). In contrast, the increase was C$0.60 per kg., or 190%, in Ching Hsien for the same period (CKNY, 1:562). In general, agricultural prices increased no more than 30% in major cities between 1910 and 1920, which was only one-sixth of the rate of increase at Ching Hsien (see Tang, p. 394 and Nankai Institute of Economics, Nankai chih-shu [Nankai price indexes] [Peking: Statistical Press, 1958], p. 12). Yenshan Hsien, another rural community some 70 miles to the northeast of Ching Hsien but less accessible to the railway, experienced general price increase of 148% in 1910-20, or some 20% less than at Ching Hsien (see Buck, "Price Changes," p. 240, table II).

[3] See Cheng Ming-ju, p. 147.

[4] Ibid. See also Maritime Customs, Decennial Report, 1922-31, Part I, p. 328.

Changch'iu Hsien, Shantung province. By 1915, peanut acreage had expanded to some 35% of total cultivated area (an increase of 35,000 times),[1] apparently consequent upon the opening of the Tsingtao-Chinan railway in 1904. In Linhsiang Hsien, Hunan province, peanuts took up about 5% of total cultivated acreage in 1915. Since the opening of the Hankow-Changsha railway in 1918, however, peanut acreage in this railway locality rapidly expanded, multiplying to 10% of total cultivated acreage by 1920 and to 20% of acreage by 1924.[2] Yet another example came from Chenliu Hsien in Honan province, which experienced a fivefold increase in the percentage of total cultivated acreage devoted to peanuts between 1915 and 1924 (from 10% to 50%) since the Lung-Hai railway was opened in 1915.[3] The importance of these examples lies in the fact that in none of the other localities studied, which were far less directly affected by railway building than the ones mentioned, was a similar magnitude of growth in peanut planting observed.[4]

The above examples can readily be multiplied. Peanuts and fresh fruits, to be sure, were hardly the only commercial crops that had benefited from rising market prices and declining transport costs. Other major industrial crops, notably soybeans, tobacco, sesame, and cotton, had all demonstrated marked expansion in acreage during the opening decades of the present century, as was mentioned before. The same economic forces that boosted peanut and fruit planting undoubtedly also left strong imprints on the spatial patterns of specialization

---

[1] Based on figures quoted in CKNY, 2:205.

[2] Ibid., 2:206.

[3] Ibid.

[4] The same study covered two other localities in North China which may be compared with Changch'iu. One is Hochien Hsien in central Hopei, where the expansion of the percentage acreage in peanuts between 1900 and 1915 was 100%. Another locality, Chiyang in Shantung, was less than 40 miles from Changch'iu. Chiyang was also about the same distance from Chinan, the major provincial market, as Changch'iu, but its accessibility to Chinan was by river, not by rail. The peanut fields in Chiyang also expanded rapidly between 1900 and 1915, measuring up to some 7,400% from 0.2% of total acreage to 15% by 1915. It can be seen, however, that both Hochien and Chiyang, despite similarity in physical and economic environments except accessibility to railways to Changch'iu, failed to exhibit as rapid a rate of expansion in peanut specialization as that which occurred in Changch'iu.
T'unghsü Hsien, which is only about 15 miles from Chenliu but farther away from the Lung-Hai line, found its peanut acreage increased from 15% of total cultivated land in 1915 to 40% in 1924. This rate of growth, though considerable, was still far less than that experienced in Chenliu which apparently received special impetus from its proximity to the newly-built railway.
A final example was from Wangpo, Hupei province, which was located some 30 miles by road from Hankow. The three localities studied in Wangpo gave peanut field expansion rates ranging from 50% to 100% in 1915-24, which were much less than Linhsiang Hsien's 300% for the same time period.

in these other crops, articulating as they were the strong influences of railways, only that regional diversity in physical conditions necessarily entailed geographical differences in the choice of product mix. This spatial affinity of high-priced commercial crops to major axes of railway construction can be illustrated by the cases of sesame and cotton in Honan, tobacco in Shantung, and cotton, peanuts, and sesame in Hopei. In all these specific cases enough empirical data are available for a closer exmination of the extent to which railways affected the degree of specialization in agricultural production.[1]

Honan was the No. 1 producer of sesame seeds in the country. In 1935, sesame output of the province was estimated at 250,000 metric tons, or about 35% of the national total. The province's estimated 5.23 million mou of sesame fields also accounted for more than 26% of the national total.[2] An official survey in 1932 revealed that half of the 36 hsien which had over 1,000 mou of farmland under sesame can be identified as railway localities (figure 10).[3] The 1932 sesame production of the railway hsien, however, was 73% of the provincial total, and their sesame acreage also accounted for some 76% of the provincial aggregate.[4] For the purpose of testing the hypothesis that railway hsien had a higher average index of specialization in sesame than non-rail hsien, sesame acreage in each of the 36 hsien was expressed as a percentage of the hsien cultivated area. The index of specialization is thus independent of the size of the administrative unit.[5] A difference-of-means test using the

---

[1] Several reasons account for the apparent lack of examples in South China. First is the absence of data. More important, however, is the scattered and small railway network in the region, which precluded the selection of a sample large enough for comparison purposes. Also relevant is the fact that certain staple foodcrops, like rice and wheat, were highly commercialized products in developed areas of South China. This necessarily makes the distinction between commercial and subsistence crops less clear in these areas.

[2] Foreign Trade Bureau, Ministry of Industry, Chih-ma, pp. 27-28.

[3] Ibid.

[4] Based on data in Forieng Trade Bureau, Minsitry of Industry, Chih-ma, pp. 27-28.

[5] There are two major reasons for using acreage in stead of production as an index of specialization. The first reason is simply the lack of local production statistics which would be needed to compute the percentage output accounted for by sesame (or other commercial crops). The second reason is that cultivated acreage, unlike output, is not affected by spatial variations in unit yields which might reflect original differences in physical production conditions. In fact, the average yield of sesame in the railway hsien in Honan was 73 catties per mou, or some 12% lower than the provincial average of 83.5 catties per mou (based on ibid., pp. 15-17). This difference in yield may imply relative inferiority in physical conditions in the railway hsien in general, but it at least suggests that ex ante productivity difference in favor of railway localities was an unlikely event.

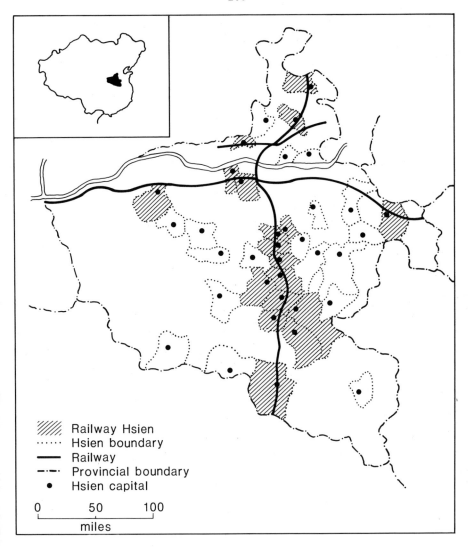

Fig. 10. Location of railway and non-railway hsien in Honan with more than 1,000 mou in sesame (c. 1932)

t-statistic was then applied.[1] The result, appearing in row (a) of table 18, rejected the null hypothesis of no difference in the mean specialization index in the two groups of hsien at the 0.05 level of confidence of a two-tailed test.[2] The much higher mean index displayed by the railway hsien can therefore be accepted as not resulting from pure chance.

Essentially the same conclusion holds for cotton cultivated in Honan, which ranked third nationally in terms of cotton acreage in the 1930s, after Hopei and Chiangsu.[3] From the statistics recorded by the Chinese Cotton Statistical Association, 43 hsien in Honan had more than 1,000 mou devoted to cotton as averaged for the years 1932-34.[4] Twenty-one of these, however, can be identified as railway localities (see figure 11). The railway hsien together accounted for 60.5% and 62.1%, respectively, of the 1932-34 average output and acreage of cotton in the province as a whole.[5] The average yield in the railway communities, however, was only 19.3 catties of cotton per mou, nearly 14% below the non-rail average of 22.4 catties per mou.[6] As indicated by the result in row (b), table 18, cotton-producing hsien along railways again exhibited a higher mean index of specialization in the crop than non-rail hsien.[7]

Shantung was perhaps the most important producer of tobacco leaves in China in the 1930s, though only about 8% of the nation's tobacco acreage was in the province, placing it after Honan and

---

[1] For two-sample tests the t-test gives essentially the same conclusions as the F-test (ANOVA), but the t-test allows inferences to be made from one-tailed tests when the direction of difference in the means can be predicted according to theory. For a brief discussion of the t-test, see A. M. Mood and F. A. Graybill, Introduction to the Theory of Statistics, 2d ed. (New York: McGraw Hill, 1963), pp. 303-6.

[2] Since the direction of difference in the means may be predicted in advance, a one-tailed test would reject the null hypothesis at a 0.025 level of confidence.

[3] See Perkins, Agricultural Development, p. 273.

[4] See Department of Agricultural Economics of the University of Nanking, Yu, Ngo, Yuan, Kan shih-sheng chih mien-ch'an yuan-hsiao, table 3.

[5] Based in data in ibid.

[6] Based on data in ibid. These were 1932-34 average yields. The 1934 average yield for the railway hsien was 22.3 catties per mou, as compared with 22.8 catties per mou for the non-rail communities as a group.

[7] Two non-railway hsien had to be dropped from the analysis because no estimates of cultivated area were available.

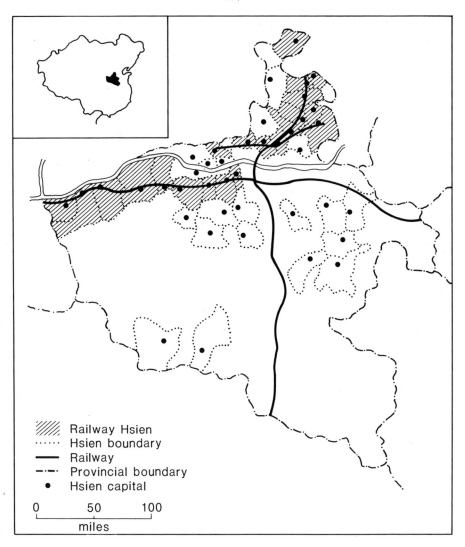

Fig. 11. Location of railway and non-railway hsien in Honan with more than 1,000 mou in cotton (c. 1932)

TABLE 18

TESTING FOR DIFFERENCES IN COMMERCIAL CROP SPECIALIZATION
IN RAILWAY AND NON-RAILWAY LOCALITIES

Province	Crop(s)	Railway Localities				Non-Railway Localities				t^{**}	Significance Level of 2-tailed test (1-tailed test)
		Number of Obser. (n_1)	% of Total Acreage	Mean* Index (x_1)	Std. Deviation (s_1)	Number of Obser. (n_2)	% of Total Acreage	Mean* Index (x_2)	Std. Deviation (s_2)		
a. Honan	Sesame	18	76%	0.1015	0.0976	18	24%	0.0411	0.0417	2.33	0.05 (0.025)
b. Honan	Cotton	21	63%	0.1193	0.1272	20	37%	0.0528	0.0457	2.14	0.05 (0.025)
c. Shantung	Tobacco	9	75%	0.0363	0.0458	24	25%	0.0031	0.0062	2.04	0.05 (0.025)
d. Hopei	Cotton, peanut, sesame	49	41%	0.1412	0.1034	59	59%	0.1065	0.1040	1.74	0.10 (0.05)

SOURCES: (a) Honan sesaem: Foreign Trade Bureau, Chih-ma (Sesame), pp. 15-17; (b) Honan cotton: Department of Agricultural Economics of the University of Nanking, Yu, Ngo, Yuan, Kan chih-sheng chih mien-ch'an yuan-hsiao, table 3; (c) Shantung tobacco: International Trade Office, Yen-yeh, pp. 16-18; (d) Hopei octton, peanut, and sesaem: Chang Hsin-i, Hopei-sheng nung-yeh kai-k'uang ku-chi (A statistical estimate of agricultural conditions in Hopei) (Nanking: Statistical Bureau of the Legislative Yuan, n.d.), pp. 32-40.

*Percentage of total cultivated area in each hsien devoted to the specific crop(s).

**$t = (\bar{x}_1 - \bar{x}_2)/s_{\bar{x}_1 - \bar{x}_2}$, where $s_{\bar{x}_1 - \bar{x}_2} = [s_1^2/(n-1) + s_2^2/(n-1)]$

Ssuch'uan.[1] The importance of Shantung in tobacco cultivation stemmed from its specialization in the production of American seed tobaccos, which were used primarily in the manufacture of cigarettes.[2] Almost all of the tobacco output in the province was marketed.[3]

According to a hsien-by-hsien survey conducted by the Ministry of Industry in 1933, tobacco took up about 400,000 mou of cultivated land in Shantung, and provincial output of the crop was a little less than 50,000 metric tons.[4] Of the 37 hsien studied by the survey, 31 had more than 100 mou under the tobacco crop, and only 9 of them were railway localities (see figure 12). The railway hsien, however, produced 55% of the tobacco output in the province, with a combined tobacco acreage amounting to 75% of the provincial aggregate.[5] A t-test confirmed that these railway hsien had a significantly higher mean index of specialization in the crop than the non-railway hsien (see row [c], table 18). But since the small size of the railway sample might have seriously weakened the normality assumption of the t-test, a nonparametric Mann-Whitney or Wilcoxon test was further applied to the rank-ordered samples.[6] The resulting U-statistic was 29, which supports the expected difference in the means at a 0.01 level of confidence of a two-tailed test (or at the 0.005 level of a one-tailed test).

A final example of distinctive specialization in industrial crops along railways draws on the combined production of cotton, peanuts, and sesame in Hopei.[7] Hopei was an important producer of

---

[1] According to an official estimate based on tax returns, Shantung's production of tobacco leaves in 1931 was 13% of the national total, or slightly higher than that of Ssuch'uan, the country's second largest producer of the crop (see Foreign Trade Bureau, Yen-yeh, p. 56).

[2] The spread of the American species in Shantung was chiefly the result of foreign entrepreneurial efforts because foreign merchants found neither the color nor the flavor of Chinese native seed tobacco suitable for the manufacture of cigarettes (see Chen Han-seng, Industrial Capital and Chinese Peasants, pp. 5-6; see also the discussion in pp. 116-17 below).

[3] This can be inferred from ibid., pp. 7, 23. In Buck's survey, 92% of the tobacco output in the Winter Wheat-Kaoliang region, which covered several tobacco-producing localities in Shantung and one in Honan (see LUSV, p. 190), was marketed. It was the highest marketing ratio reported among the eight agricultural regions studied as regarding tobacco.

[4] Foreign Trade Bureau, Yen-yeh, p. 19.

[5] Based on data in ibid., pp. 16-19.

[6] For a brief discussion of the Rank Test, see Mood and Graybill, pp. 417-19.

[7] Soybean has been excluded in the analysis since only about 26% of soybean output in the Winter Wheat-Kaolian region was usually marketed (see Buck, LUSV, p. 228).

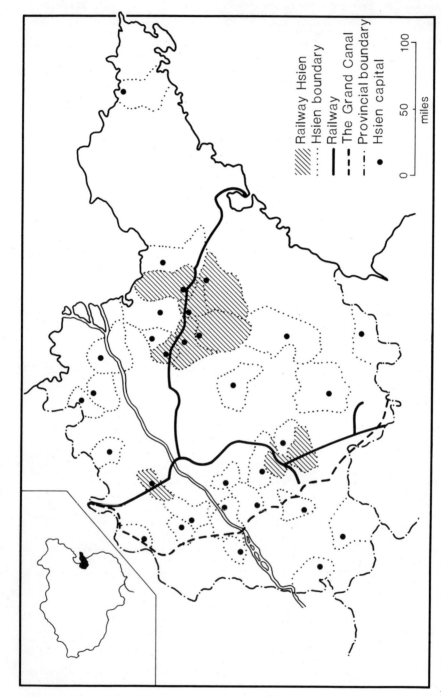

Fig. 12. Location of railway and non-railway hsien in Shantung with more than 100 mou in tobacco (c. 1933)

all three crops and probably ranked second in the country both in
terms of acreage and output in all three cases in the 1930s.[1] The
present analysis is based on a 1929-31 agricultural survey carried
out by the Statistical Bureau of the Legislative Chamber (Yuan) of
the Nationalist government.[2] According to this survey, only 21 of
the 129 hsien in the province had less than 1,000 mou of cultivated
land occupied by the cotton crop. Most of these cotton-producing
localities were also major peanut producers and, in fact, only three
localities in which peanut acreage was more than 1,000 mou did not
also belong to the cotton group. On the other hand, only 15 hsien
were recorded as having more than 1,000 mou devoted to sesame
production, and all of these hsien simultaneuosly had more than 1,000
mou of land under the cotton crop. Total sesame acreage accounted
for no more than 0.25% of the total cultivated area in the province.

All hsien with more than 5,000 combined mou engaged in the
production of cotton, peanuts, and sesame were included in the
present analysis. Of the resulting 108 hsien thus selected, 49 were
identified as railway localities (figure 13). The railway hsien had
a combined cotton, peanut, and sesame acreage amounting to about 4.6
million mou, or 41% of the provincial total, but their aggregate
output of the three crops was only some 36.8% of the provincial
figure. The obvious explanation for this lies in the generally lower
yields per unit of cultivated land in the railway localities for
both cotton and sesame, though peanut yield on the average was
somewhat higher than that realized in the province as a whole.[3] A
t-test again was applied to the mean indexes of specialization in the
railway and non-railway samples. The result, given in row (d) of
table 18 (p. 108), barely accepted the null hypothesis of no
difference in the two groups at the 0.05 level of confidence of a
two-tailed test. However, since the direction of the difference was
predicted at the outset, the higher index pertaining to the railway

---

[1] For cotton statistics see Perkins, Agricultural Development, p. 259. For peanut output and acreage estimates see Foreign Trade Burea, Hua-sheng, pp. 27-31. For sesame statistics see Foreign Trade Bureau, Chih-ma, pp. 27-28.

[2] Chang Hsin-i, Hopei-sheng nung-yeh kai-k'uang ku-chi (A statistical estimate of agricultural conditions in Hopei)(Nanking: Statistical Bureau, n.d.).

[3] Cotton and sesame yields in the railway localities were averaged at 26 and 67 catties per mou, respectively, as compared with 28 and 69 catties per mou for cotton and sesame, respectively, in the province as a whole. Peanut yield in the railway areas came to about 239 catties per mou on the average, as compared with the provincial average of 230 catties per mou (based on data in ibid., pp. 32-40). However, some other estimates from official surveys gave 236 to 238 catties per mou as the provincial average for peanuts (see Perkins, Agricultural Development, p. 282).

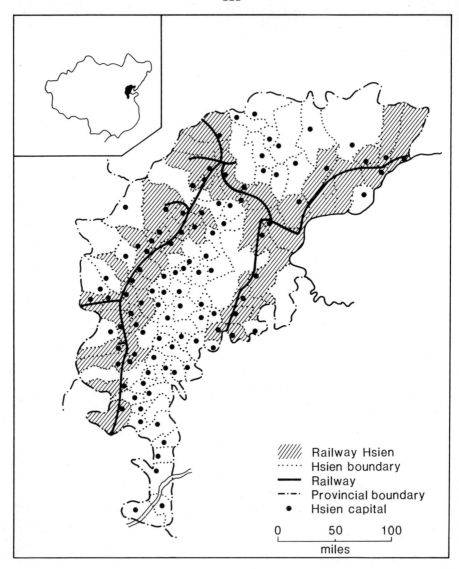

Fig. 13. Location of railway and non-railway hsien in Hopei with more than 5,000 mou in cotton, peanuts, and sesame (c. 1930)

sample may be accepted as significantly different from that displayed by the non-railway areas at the 0.05 level of a one-tailed test.

The above empirical case studies clearly indicate that the pattern of commercial crop specialization did appear to be more pronounced along railway lines in prewar China. The development of rail transport not only brought higher farm prices to agricultural producers along railways, it also effectively introduced special opportunities for scale economies in production and exchange because of the availability of fast, regular, and reliable transportation service that tied local economies to non-local sources of supply and demand. A greater degree of specialization in non-food industrial crops meant less farmland available for food production and thus heavier dependence on the market for food supply on the part of farmers. In areas where fast transportation links with potential food surplus regions were unavailable, the extent of specialization inevitably measures the risk to be borne by cultivators who helplessly faced erratic seasonal fluctuations in food prices and possibly in crop yields. Railways, however, enabled local economies to economize on the stock of food previously required to maintain a given standard of living by promoting interdependence of regional markets. Though nowhere in the above empirical cases were the mean indexes of specialization impressively high (they never exceeded 15% of cultivated area on the average), many localities, especially, though not exclusively, of the railway group, were in fact characterized by index values well above 25%.[1] The above calculations also neglected many other commercial and semi-commercial crops that were actually planted. For instance, if it is assumed that a mere 26% of the wheat and 24% of the soybean acreage were given to production for the market,[2] the inclusion of such acreages in the calculation would have raised the average index of the railway localities in Hopei from 0.14 to 0.26, with over 20% of these sample localities actually having more than 35% of the farmland devoted to the cash crops. Moreover, the empirical evidence studied above pertained to a period of steady

---

[1] There were two hsien, both railway localities, with more than 25% in index values in the Honan sesame case. In the Honan cotton case, there were three such hsien, all railway localities. For the Hopei combined crop case, a total of 17 localities had more than 25% in index value, 11 of which were railway localities.

[2] These figures refer to the average percentage of the specific crop being marketed in the Winter Wheat-Kaoliang Region reported in Buck's survey (see LUSV, pp. 227-28). In fact, a much higher figure may be expected for areas along railway lines (see below).

or declining relative prices for farm products.[1] In periods of rising cash crop prices, much greater specialization would be observed. During the late 1910s and early 1920s when cotton prices were rising, it was observed that, in Hopei, in the railway communities of "Chengting, Tingchou, Hsinlo, and in the neighboring <u>hsien</u> of Shihchiachuang, . . . cotton planting greatly flourished . . . Cotton acreage took up as much as 70-80% of all cultivated land, so that food imports from Shanhsi had to be relied upon to make up for local food shortages."[2] Of particular importance is the spread mechanism of agricultural commercialization here elicited: increased specialization in industrial crops in regions of improved transport conditions simultaneously stimulated foodgrain marketing and production in less locationally advantaged areas. Rural income improvement in regions along axes of transport development is therefore diffused, through the generation and intensification of interregional trade, to otherwise subsistence regions as well.

However, the patterns of crop specialization still inadequately reflected the actual extent of market participation of farmers. The approach leaves out of account the actual variability in the commercial nature of the same bundle of farm products, depending as it does on the relative accessibility to markets. Thus capitalizing on improved transport conditions, agricultural producers with easy access to rail-heads not only tended to enhance their specialization in marketable products, but they actually marketed a higher proportion of the output than their counterparts elsewhere. A 1934 survey of rural communities along the Peking-Hankow railway revealed that an average of 85% of the cotton, 75% of the soybean, and 97% of the sesame production were marketed in the 30 <u>hsien</u> studied in Honan and Hopei.[3] Wheat, a largely subsistence crop in more isolated communities in North China, assumed a definite semi-commercial status

---

[1] The agricultural price index in North China fell nearly 40% between 1930 and 1934, as compared with 11% for manufactured products and 13% for imports as a group (see Nankai Institute of Economics, <u>Nan-kai chih-shu tzu-liao hui-p'ien</u>, p. 13, for general agricultural and manufactured product price indices; see Hou, p. 232, for adjusted Nankai indices). A similar magnitude of deline in the relative prices of agricultural products in South China was also observed (see Shanghai Economic Research Office of the Chinese Institute of Sciences, <u>Shanghai chieh-fang ch'ien-hou wu-chia tzu-liao hui-p'ien</u> [A compilation of price materials for pre- and post-Liberation Shanghai][Shanghai: Peoples' Press, 1958], p. 135).

[2] <u>CKNY</u>, 2:212; see also ibid., 2:133.

[3] See Ch'en Pei-chuang, <u>Ping-Han t'ieh-lu yen-hsien nung-ts'un ching-chi tiao-ch'a</u> (An economic survey of rural villages along the Peking-Hankow Railway) (Shanghai: Chiao-t'ung University Press, 1936), Appendix, table 16. The several localities in Hupei have been excluded from this calculation to allow comparison with Buck's figures.

in these railway localities with an average of 55% of the harvest crop marketed.[1] An independent study by Gamble on 400 farms in Ting Hsien, Hopei province, on the Peking-Hankow line in 1927 also found that 86% of the cotton, 93% of the peanuts, 92% of the sesame, and 42% of the wheat output were marketed.[2] These figures were substantially higher than those reported by Buck for the Winter Wheat-Kaoliang Region as a whole (cotton, 37%; soybean, 24%; sesame, 76%; wheat, 26%),[3] though Buck's survey is generally regarded as already biased towards favorably located areas. On the other hand, in regions where the comparative advantage in foodgrain production was greatly reinforced by railway development, even foodgrains were suddenly found to become the mainstay of cash output. In communities along the Peking-Suiyuan railway, for instance, it was reported that wholly four-fifths of the local grain output were exported to deficit regions along other railway lines.[4] Since changes in agricultural prices affected cultivators' incomes in proportion to their sales of produce, regional economies efficiently served by rail transport benefited increasingly more from rising market demands for agricultural goods associated with industrial and urban developments.[5]

Though the greatest impetus to commercialization along railway lines clearly derived from economic incentives associated with improvement in transport conditions, a number of institutional factors, both endogenous and exogenous, might have contributed directly or indirectly to the commercialization process. One of these was the increased monetization of local economies where commercial cropping rapidly expanded. Land taxes, which rose accordingly if laggardly with improvement in land productivity, had to be paid in

---

[1] Ibid., Appendix, table 9. The figure refers to owner-operator output only, but output of tenant farmers was small.

[2] Quoted in Myers, "The Commercialization of Agriculture in Modern China," p. 184.

[3] LUSV, pp. 227-28.

[4] See CKNY, 2:233; see also ibid., 2:134.

[5] Obviously the higher degree of market dependency also meant greater reduction in incomes for the local economies in periods of general recession and declining agricultural prices (as compared with non-railway areas). But if the relative decrease in farm prices was greater for the railway localities, the absolute decrease in farm prices would be the same for all areas alike. The original difference in farm price levels in railway and non-railway areas would be unaffected, and railway localities still had higher incentives to produce cash crops than non-railway localities and thus maintained a higher level of incomes.

cash.[1] A similar but obviously more sensitive relation existed between land productivity and land rents.[2] Cash rent collection, as contrasted with in-kind rental payments, was largely practiced in areas with relatively high levels of specialization in commercial crops.[3] The increased demand for cash income for tax and rent payments therefore added to the cash demand in connection with the increased purchases of food, productive inputs, and imported consumer goods as farm income improved. A self-reinforcing mechanism was thus established in the commercialization process itself.

Private entrepreneurial efforts was another factor that promoted commercial cropping along railways. The most notable case was the pioneering attempt of the British-American Tobacco Company (B.A.T.) and the Nanyang Brothers Tobacco Company, a Chinese concern, at introducing and promoting the cultivation of American seed tobacco in selected, mainly rail-adjoining, localities in Shantung, Honan, and Anhuei during the early and mid-1910s.[4] In order to dispel the suspicions of native cultivators, the B.A.T. and its compradores offered incentives ranging from the free distribution of seeds and fertilizers to cash payments for entire crop, irrespective of quality, at best prices.[5] These strategies paid off swiftly and handsomely as local peasants flocked to enter into the tobacco-planting business.[6] Following similar efforts by the Nanyang Brothers, tobacco acreage rapidly expanded at the original experimental sites and cultivation of the crop steadily diffused to many other districts on or close to the Ch'ingtao-Chinan and

---

[1] In Republican China, the rapidly growing land tax burden, due in particular to the levy of a great multitude of local surtaxes from military requisitions to construction and administration fees, might actually have been an exogenous factor in the increased commercial production of hard-pressed farmers. To what extent this was indeed the case is hard to tell, but peasants in more market-accessible areas would generally be better off than their counterparts elsewhere because of the higher prices they could receive for their sales.

[2] See, for example, Chen Han-seng, pp. 80-85.

[3] See Ch'en P'ei-ta, Chin-tai chung-kuo ti-tso kai-shuo (Introduction to land rents in modern China), 2d ed. (Peking: People's Press, 1963), p. 26.

[4] The major experimental sites of early years were Wei Hsien (Shantung), Fengyang (Anhuei), and Hsiangch'eng (Honan). They were chosen both because of the suitability of the soils and their accessibility to railway lines (see Chen Han-seng, pp. 6-7).

[5] See Chunjen C. Chen, "Tobacco-Growing in Shantung," Chinese Economic Journal 10 (January 1932):37-44; Chen Han-seng, pp. 6-7; G. C. Allen and A. G. Donnithorne, Western Enterprise in Far Eastern Economic Development (New York: Kelley, 1968), p. 171; and W. Y. Swen, "Types of Farming, Costs of Production, and Annual Labor Distribution in Wei-hsien, Shantung," Chinese Economic Journal 3 (July 1928):658.

[6] See Allen and Donnithorne, p. 170.

Peking-Hankow (Honan section) lines, along which numerous curing factories and collection stations were also established.[1]

The choice of railway sites as experimental stations was hardly an arbitrary decision. The convenience of rail transport and the many economic advantages associated with railway sites were as well recognized by private business concerns as by public or semi-official agencies interested in promoting the production of specific imported species of commercial crops. Both Japanese merchants and the Chinese Cotton Improvement Commission, for example, expressly labored to promote the cultivation of American cotton in areas along railway lines in North China.[2] Many government-sponsored experimental stations, too, were established at railway sites.[3] Whereas the convenience of transport was clearly one major factor, the attraction of railway localities probably also lay in the greater readiness of local peasants to experiment with total unfamiliar varieties of species due to greater market incentives in more accessible areas. But even without conscious promotional efforts, it could only be natural for imported species to be first adopted along railway lines because initial channels of spatial diffusion were invariably routes of modern communications. Foreign peanut species, for instance, were reportedly imported via the Peking-Hankow and Loyang-K'aifeng railways into the inland territories soon after the completion of the lines.[4] Accessibility to railways was thus important in securing the supply of new, improved-quality species from abroad and in improving or even maintaining product quality because of undesirable crossings of imported species with low-quality native species over time.[5]

Perhaps an outcome bred from the interaction of economic and institutional factors, the appeal of railway localities was also reflected in the distribution of institutional agricultural credits. The organization of commercial bank-backed credit cooperatives and the destination of agricultural loans distributed as working capital by commercial banks were both highly in favor of railway communities or, in general, localities highly accessible to modern channels of

---

[1] See Chen Han-seng, pp. 16-23; CKNY, 2:225-26; and Allen and Donnithorne, p. 171.

[2] See, for example, CKNY, 2:156, 157, and 3:169.

[3] Ibid., 2:158.

[4] Ibid., 2:133.

[5] See, for example, Department of Agricultural Economics of the Nanking University, Yu, Ngo, Yuan, Kan shih-sheng chih mien-ch'an yuan-hsiao, p. 33.

communications.¹ In view of the lower risk involved and the greater demand for capital in railway communities because of the greater specialization in commercial planting, this pattern of commercial loan disbursement may indeed be expected. But the availability of agricultural credit was such a potent factor in the development of traditional agriculture that its discrimination against transportationally disadvantaged regions could only aggravate imbalanced growth in the rural economy. The implication for agricultural growth in the national economy was particularly grave when, as in prewar China, apparent neglect on the part of central and local governments had caused a low level of development in local and intraregional transport infrastructures.

Marketable Surplus and the Effect of Railways

The "marketable surplus" of a crop partly grown for subsistence purposes is the total farm output of the crop minus the amount of on-farm consumption by the peasant household. "Marketable surplus" is sometimes measured conveniently, if data exist, in terms of the "marketed quantity" of the crop output, provided that the latter refers to the net amount of output sold for cash; i.e., it is net of the quantity later bought back by the peasant household for consumption purposes because of earlier forced sales. In discussing the "marketable surplus" of total farm output of smallholder farms as in the present study, the concept is taken to mean the net value of farm output that can be marketed after home consumption needs are met. By expressing the "surplus" in value terms the concept can readily take into account the substitutability among categories of farm products actually consumed, including products from animal husbandry.

The size of the marketable surplus produced by a peasant household is a convenient measure of agricultural commercialization at the farm level. It in effect also gauges the potential capacity of the farm operator to invest in agricultural production, which affects farm productivity and future incomes of the household. When part of the cash income is spent on manufactured goods, its aggregate impact is likely to benefit domestic industrial growth and the development of import-substituting industries besides facilitating capital formation in the farm sector.² But what is most emphasized by development theorists is the basic role of a "food surplus" in the

---

¹See CKNY, 3:183, 184, and 187.

²See note 2 on p. 98 above.

long-term growth of a traditional agrarian economy: the "surplus" on which urbanization, industrial development, and, by implication, improvement in agricultural productivity so heavily depend. A common recognition in the development literature is that any policy or set of policies that leads to expansion in the marketable surplus also necessarily contributes to the pace of national economic growth.[1]

As can be easily understood, the marketable surplus is intimately related to the productivity level in the farm economy, which in turn is directly affected, other things being equal, by the degree of specialization in agricultural production. For centuries there has been a high degree of specialization in agrarian China, and to most Chinese peasant farmers marketing produce was just as important a part of life as sowing and reaping. To exactly what extent Chinese farmers had participated in a market economy in pre-modern times cannot be ascertained, but the steady evolution of central-place systems, especially the rapid multiplication of market towns since the seventeenth century and the later growth of large urban centers could provide some idea of the requisite "surplus."[2] By the first half of the nineteenth century non-farm population had grown to account for about 8% of total population,[3] and it was likely that 10-20% of the gross agricultural output of the country was consumed outside the farm sector;[4] but it was not until the late nineteenth and early twentieth centuries that a real revolution was to take place. By the early 1930s, probably over 20% of the country's population can be classified as non-agricultural,[5] and as much as

---

[1] See, for example, William H. Nicholls, "Development in Agrarian Economies: The Role of Agricultural Surplus, Population Pressures, and Systems of Land Tenure," Journal of Political Economy 71 (February 1963):1-29; J. M. Hornby, "Investment and Trade Policy in the Dual Economy," Economic Journal 78 (1968): 96-107; A. K. Dixit, "Marketable Surplus and Growth in the Dual Economy," Journal of Economic Theory 2 (1970):107-21.

[2] There is evidence that market towns accelerated their growth and approached a rate even faster than population growth in the seventeenth century (see Elvin, p. 268).

[3] This is based on the estimation by William Skinner for 1843 (see William Skinner, "Regional Urbanization in Nineteenth-Century China," in idem, ed., The City in Late Imperial China [Stanford: Stanford University Press, 1977], p. 229, table 4). I have assumed the same rate of population growth in central places with population under 2,000 as that in towns with 2,000 people or more between 1843 and 1893.

[4] Apparently if one assumes the same rate of per capita consumption in the non-farm population as in the farm population, non-farm consumption of the agricultural output would be 8% of the gross national total, plus 1-2% for exports. Higher incomes in the non-farm sector and the raw material demand of handicraft industries would of course mean the actual "marketed surplus" was higher than 10%.

[5] See Buck, LUSV, p. 420.

30-40% of the gross agricultural output might be marketed to non-farm consumers.[1]

In general, despite the 1-2% per annum growth rate in rural population and the 3-5% per annum growth rate in urban population during the half century prior to the Second World War, China's food production had been able to match the growth in population while industrial crop production also rapidly expanded.[2] Since neither the increase in cultivable acreage nor that in foreign imports is likely to have contributed appreciably to the results,[3] the sizable growth in marketable surplus since the late nineteenth century in prewar China could only be attributed to rises in farm productivity and increased agricultural specialization. In this perspective the role of railway development in the expansion of marketable surplus becomes apparent.

An exposition of the relationship can, again, be sought via the impact of railway development on the market accessibility of agricultural producers. Since marketable surplus is, by definition, the difference of two components, production and consumption, the effect of accessibility on marketable surplus can be understood as the sum of its separate effect on production and consumption. The effect of an improvement in accessibility on agricultural production, as analyzed in the previous chapter, has been demonstrated to be positive. The effect of accessibility increase on consumption, however, is also likely to be positive. Since improvement in accessibility improves farm incomes, farm consumption levels would rise as long as the income elasticity of demand for farm output, especially foodstuffs, is positive. Nevertheless it is unlikely that farm households would tend to consume the entire increase in

---

[1] Several rural surveys in the 1920s and 1930s gave estimates of marketed agricultural produce ranging from 38-58% of farm output (in value terms) (see CKNY, 2:271, 269; 3:312, 315). Buck, on the other hand, reported that on the average only 12% of the marketed output was sold to farmers (LUSV, p. 343). If we take 40% as the average of farm output marketed, then about 35% of total output may have been sold to non-farm consumers.

[2] See Myers, "Commercialization of Agriculture," pp. 174, 182.

[3] By the nineteenth century, China had begun to run out of readily cultivable land. Most of the increase in cultivated land in the late nineteenth and early twentieth centuries went into low-quality land in Manchuria, Inner Mongolia, and the northwest, which in fact may have caused the slow decline in average yields of some food staples since the 1800s (see Perkins, Agricultural Development, pp. 27-28). As for foreign food imports, they remained insignificant as a source of supply for domestic consumption needs. Grain imports, for example, never exceeded 0.2% of domestic consumption levels in the prewar era (see Cheng Yu-kwei, Foreign Trade, p. 257, note 62). A study by Friedrich Otte also showed that coastal urban centers became dependent on imports of cereals only when the grain supply from the countryside was interrupted by poor harvests or by civil war (quoted in Myers, "Commercialization," p. 1).

output, or to spend all incremental cash incomes on farm sector products. For this reason it may be reasonable to expect that a positive relationship exists between accessibility and the average marketable surplus of farm families.

Whether this relationship actually existed in prewar China can be examined empirically with the detailed consumption data collected in Buck's survey. The survey provided in detail the average annual consumption (in grams) per person of major categories of food in farm households of 136 hsien.[1] It also gives the average percentage of each category of food consumed which was supplied by the farm itself. The survey, however, did not report the average per person consumption of non-food crops in farm households; but since in most cases the crop acreage devoted to non-food industrial crops (mainly cotton, tobacco, and hemp) was small, exclusion of these crops in the analysis would not seriously affect the results.[2] To estimate farm consumption, the average daily consumption per person of all major categories of foodstuffs in a locality was converted into kg. wheat or rice equivalents, according as the locality is in the Wheat or Rice Region, by means of relative prices.[3] Processed foodstuffs, such as flour and vegetable oils, were included, but only that portion which was supplied by the farm itself was calculated. Total annual consumption per person was daily consumption per person times 365, and annual consumption per farm household was total annual consumption per person times the average number of persons per farm household in a given locality. Average farm output, on the other hand, was expected farm output times survey year yield as a percentage of the most frequent yield in a locality, which is given in Buck.[4] Estimated marketable surplus per farm was given by the difference between realized production and actual consumption in the farm, net of the average tax payment per farm in a given locality.[5]

---

[1] LUSV, pp. 86-121.

[2] For the survey sample as a whole, cotton occupied only 3.4% of the aggregate cropped area, hemp, 0.2%, and tobacco, 0.8% (see LUSV, p. 176 and p. 178).

[3] As in the estimation of farm output (appendix A, pp. 147-48), provincial average prices were used. Where such prices cannot be obtained, approximate ratios of nutritional values of the foodstuffs were used as surrogates (nutritional values are given in LUSV, p. 67).

[4] LUSV, p. 208.

[5] The average tax paid to the government per hectare of the most usual kind of farmland in a locality in the year in which the consumption survey was carried out is given in Buck, LUSV, p. 165. I have assumed here that the average farm was owner-operated since only about 5% of all farms in the survey was worked by tenant farmers (see LUSV, p. 57).

The statistical relationship between marketable surplus per farm and accessibility is given by the empirical estimation of equation (1) using Ordinary Least Squares:

$$\ln M = a + b \ln A \qquad (1)$$

where M = marketable surplus per farm in kg. grain (wheat or rice) equivalents; A = accessibility index; b = elasticity of marketable surplus with respect to accessibility; and a = constant intercept. The estimation results of equation (1) for the Wheat and Rice Regions appear in table 19. As anticipated, marketable surplus increases as accessibility improves. The degree of the statistical association is about the same in the Wheat as in the Rice Region: each 1% increase in accessibility (or a 1% decrease in transfer costs) was associated with about 0.38% increase in marketable surplus.

TABLE 19

REGRESSIONS RELATING MARKETABLE SURPLUS TO ACCESSIBILITY

Region	Constant	Slope	F-Ratio	R2	Sample Size
Wheat	8.474	0.388	37.11*	0.378	61
Rice	8.711	0.383	28.23*	0.296	67

*Significant at 0.01 level of confidence.

In order to see if railway localities again displayed a distinctly higher level of marketable surplus than other localities, mainly a result of the greater incentives to produce for the market and to specialize, an ANOVA test similar to the one given in table 16 was applied to the dichotomized rail vs. non-rail data. Because the consumption survey gave data on 136 hsien instead of 168 hsien as was the case in the production survey, the resulting sizes of the railway samples are 27 in the Wheat Region and only 11 in the Rice Region, and those of the non-rail samples are 22 in the Wheat Region and 40 in the Rice Region. Despite the slight reduction in the sample sizes, the ANOVA test confirmed that in both the Wheat and Rice Regions, railway hsien did exhibit a uniformly higher level of marketable surplus than non-railway hsien (table 20). This result evidently supported the argument that railway development, by improving the opportunity for profitable exchange, also increased the potential capacity of local farm households to invest in productive activities because of the increases in marketable surpluses. The outcome, in effect, was a relative improvement in the expected future incomes of farming families in areas directly affected by railway construction. The increase in marketable surplus, of course, would also mean that railway areas in general contributed a larger share of the aggregate surplus to the non-farm sector than non-rail areas. This, apparently, was also demonstrated by the greater specialization in railway areas as examined earlier.

TABLE 20

ANOVA TESTS OF RAILWAY EFFECTS ON MARKETABLE SURPLUS

Region	Mean Surplus: Railway Hsiens	Mean Surplus: Non-Rail Hsiens	Computed $F^{\#}$ ($n_1=1$, $n_2=d$)
Wheat	1,789	1,301	6.23* ($n_2=47$)
Rice	3,266	2,659	9.11** ($n_2=49$)

$^{\#}n_1$ and n_2 denote between-group and within-group degrees of freedom, respectively.

*Significant at 0.05 level of confidence.

**Significant at 0.01 level of confidence.

Summary

This chapter has been devoted to an examination of the impact that rail transport development had on agricultural commercialization at the farm level in prewar China. Agricultural commercialization measures the extent to which the productive efforts of farm operators were directed at earning a money income. Given data constraints, the degree of commercialization in local farm economies can only be inferred from circumstantial evidence such as the level of farm specialization in commercial crops. Case evidence from prewar North China enables the direct statistical testing of commercial crop specialization in railway versus non-railway areas in a number of provinces. In all cases examined the railway communities generally exhibited a higher level of specialization than non-railway communities. When other missing factors, such as the cultivation of semi-commercial crops, the greater commercial nature of the same crop product in railway than in non-railway areas, the deteriorating terms of trade for farm products in the years in which the data were collected, are taken into consideration, the empirical evidence clearly indicated that the commercialization impetus afforded by railways to the farm economy was considerable. A number of institutional factors, however, also tended to reinforce the momentum towards commercialization received by railway localities. These included the monetization process that fed on the commercialization process itself; the efforts by private and public agencies to promote imported crop species cultivation along railway lines; the tendency for imported crop species to diffuse along modern channels of communications; and the concentration of institutional agricultural credits in highly accessible regions.

Even stronger implications for the economic development of the agrarian economy may be revealed when the impact of railway development is seen in terms of the marketable surplus of farm households. The marketable surplus of a farm, defined as the surplus of output over household consumption, signifies the potential capacity of the farm household to invest in productive farm activities or assets. It, too, when considered in the aggregate, reflects the productivity of the farm sector which directly affects the rate of urbanization, domestic industrial development, and the long-term growth of the national economy. Because of the detailed consumption statistics collected in Buck's survey, it was possible to estimate the average marketable surplus of farm households in sampled localities. As in the analysis of farm output in the previous chapter, the effect of railway development on marketable surplus has been examined both via the relationship between marketable surplus and accessibility, and in terms of the comparative level of marketable surplus in railway and non-railway localities. Both findings, as expected, gave firm support to a positive influence of railway development on the size of marketable surpluses. The importance of this conclusion lies, perhaps, not so much in the economic opportunities accorded a small segment of China's peasantry by an unimposing system of railways as in the pivotal role of transport development, as exemplified by the case of rail transport, in the economic growth of traditional economies.

CHAPTER VI

TRANSPORT DEVELOPMENT AND RURAL ECONOMIC GROWTH:

A GENERAL ASSESSMENT

Introduction

The analyses carried out in the preceeding chapters have been geared to answering the question: "Has the development of rail transport in China exerted any appreciable influence of a positive nature on agricultural production activities and hence on agricultural incomes?" The empirical evidence we have examined apparently gave this question an affirmative answer. However, the change in the income position of farmers and their households in an agrarian economy cannot be fully understood without a clear understanding of the impact which the same economic forces that affected agricultural output might have on other non-farming production activities regularly pursued by the farm households. These non-agricultural activities, whose products are conveniently termed "Z-goods" following a conventional terminology, include non-resident rural service industries such as the provision of transportation and home handicraft industries such as the spinning and weaving of cloth, the processing of raw foodstuffs, and so forth. As late as the early 1930s the output of these Z-activities probably still accounted for as much as 11 per cent of the Gross Domestic Product in China, or about 17 per cent of the gross value added of the domestic agricultural output.[1]

The objective of this study has been to achieve an understanding of the economic impact of railway development on agriculture, and as such it could have easily dispensed with the burden of bringing the Z-activities into the picture. But given the economic organization of production in agrarian China in which resources were to be allocated between agricultural and Z-activities,[2] the output of, and the incomes generated from, both types of rural activities are closely interrelated. The treatment of the Z-activities can thus be

---

[1]Derived from Liu and Yeh, p. 66, table 8. I have combined old-fashioned transportation and the handicraft industries to arrive at a gross total of the net output of the Z-activities.

[2]Some scholars prefer to treat Z-activities as the domain of surplus labor so that they are not substitutable with farming. (See, for instance, Kang Chao,

viewed as an extension of previous analyses and, when conceptualized in an integrated theoretical framework, can be handled quite easily by simply enlarging the interpretive scope of the present results. The expanded framework also sheds light on a controversial thesis maintained by many social theorists on the economic history of modern China: that foreign investment and development of the modern sector, including modern advances in the transport sector, actually impaired, rather than benefited, the rural economy. If this were true, railway development could actually have brought about a net social loss to the country because of the sheer size of the rural sector.

In order to integrate the results of this study into a theoretical framework which incorporates nonagricultural activities, I have drawn on the Hymer-Resnick model.[1] A brief exposition of the model and its implications for rural development in China will be undertaken in the ensuing section. However, even at this point, with the recognition that rail transport development has indeed contributed to the economic progress of agrarian China, the present study failed to give a fair assessment of the potential contribution of transport advances to the traditional economy. Some major causes of the divergence of potentiality and reality in this respect will thus be examined, and particular attention will be given to the role of the government in promoting or undermining such causes.

Trade Development and Rural Growth: Theory and Evidence

As is characteristic of semi-subsistence agricultural economies, the economic organization of family farms in traditional China revolves around two types of production activities: agricultural and nonagricultural. The output of the latter type of activities, termed "Z-goods" for convenience, is predominantly for home consumption purposes and can practically be considered as non-traded commodities, especially when one takes the local community as a single production

---

The Development of Cotton Textile Production in China [Cambridge, Mass.: Harvard University Press, 1977], chapter 7.) The concept of a pool of "surplus labor" that existed on an annual basis, however, is difficult to accept empirically because of the large amount of evidence on the acute shortage of farm labor at busy seasons (see, for instance, Buck, Land Utilization, pp. 299-301). Even the view that surplus labor existed in slack seasons may have neglected the usually considerable flexibility in the choice of techniques available to even the most primitive agrarian economy (see Stephen Hymer and Stephen Resnick, "A Model of An Agrarian Economy with Nonagricultural Activities," American Economic Review 59 [Sept. 1969]:494-95).

[1] Ibid., pp. 493-506.

unit. On the consumption side a third category of commodities enters into the picture: manufactured imports from the urban-industrial (or modern) sector and from abroad which are close substitutes for the Z-goods. Transport development has improved the competitiveness and availability of the manufactured goods (M-goods) in rural China since the 1890s.

The interrelationships between the consumption and production aspects of a rural community in prewar China in face of development in the modern sector are portrayed by the model in figure 14. The concave curve AB in quadrant II summarizes the production possibility combinations of Z and F (agricultural output) as resources are efficiently allocated between their production.[1] The terms of trade between F and M, where $P = P_M / P_M$, is given by the price lines in the third quadrant. Suppose all F produced is sold on the open market to exchange for M at a given P,[2] then points can be chosen on the price line which, in combination with corresponding points in the second quadrant, provide the consumption possibilities schedule in the first quadrant, denoted by the curve ACQ. At a prevailing price P, agricultural output amounting to OK thus corresponds to a consumption equilibrium position at C, where the consumption possibility curve is tangent to the community indifference curve $U(Z,M)$. At this position all OK units of F produced would be exchanged for OG units of M at the given price P, and OI units of home-produced Z would be consumed.

An increase in the relative price of F, say from P to P', shifts the consumption possibilities curve to ADR and the consumption equilibrium point to D in a comparative static framework. This situation is attained when agricultural output is expanded from OK to OL, which is then exchanged at price P' for OH of M. At the final equilibrium position denoted by D, the amount of Z goods produced has dwindled from the prior OI units to OJ units, and M consumed has increased by GH. Since D is on a new community indifference curve $U'(Z,M)$ which has a higher ordinal ranking than $U(Z,M)$, the rural community at D is better off than it was at C.

An important recognition concerns the shape of the offer curve ACD, which is obtained by varying P and which traces out the

---

[1] The concavity of the production possibility curve between F and M derives from the usual assumption of the diminishing marginal rate of technical transformation between F and M. This assumption, however, is not necessary for the functioning of the present model.

[2] Allowing some F to be consumed at home, a more realistic assumption, complicates the model but the main conclusions of the model are unaffected (see Hymer and Resnick, p. 499).

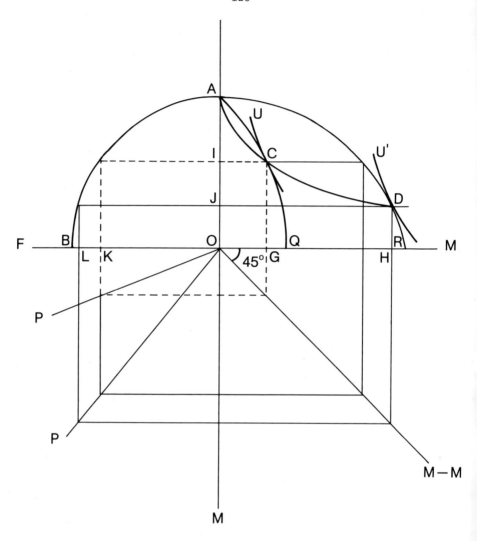

Fig. 14. A model of production and consumption in an agrarian economy with nonargicultural activities (after Stephen Hymer, "The Decline of Rural Industry under Export Expansion," Jour. Econ. History 30 [March 1970]: 54)

equilibrium consumption positions C, D, etc. If Z is a normal good (i.e. one which has a positive income effect), increases in F marketed would lead to increases in farm incomes and thus a tendency to increase the consumption of Z. Rising agricultural price (P), which at first induces the increased production and marketing of F, automatically boosts farm incomes and the consumption of Z, and thus eventually helps to reduce the supply of F. This happens when the income effect of rising P overwhelms the substitution effect associated with the deteriorating relative price of M with respect to Z. As a result the offer curve ACD would bend upward and backward, with a corresponding drop in the supply elasticity of F.[1] However, if Z is an inferior good so that the income effect becomes negative, further rises in P actually would reinforce the substitution effect, and the supply elasticity of F would keep on increasing until complete specialization in F is achieved.

The characterization of Z as an inferior good is, in fact, supported by historical evidence from a number of developing agrarian economies studied by Hymer.[2] The case of prewar China probably presents no exception. The gradual displacement of a variety of handicraft industries by modern manufactures in prewar China is a well-documented issue which has been seized upon by many as clear indication of the disruption that developments in the modern sector have brought to the traditional, rural economy.[3] However, the decline of rural industries in China occurred at a time when the terms of trade turned increasingly in favor of farm exportables, a result which, as mentioned before, may be attributed not only to urban and industrial developments and growth in foreign demand but also to transport development in the countryside. Increasing agricultural commercialization and improvement in agricultural incomes hence

---

[1] For an algebraic exposition of the model and its mechanisms see Hymer and Resnick.

[2] Stephen Hymer, "The Decline of Rural Industry Under Export Expansion: A Comparison among Burma, Philippines, and Thailand, 1870-1938," Journal of Economic History 30 (March 1970):51-73.

[3] Strong expression of this view can be found, for example, in Yen Chung-p'ing, Chung-kuo mien-fang-chih shih-kao [Draft history of Chinese textiles] (Peking: K'o-hsüeh ch'u-pan-she, 1955); Fei Hsiao-tung, Peasant Life in China (London: Routledge and Kegan Paul, 1939); idem, Earthbound China; R. H. Tawney, ed., Agrarian China; and Fang Hsien-ting (H. D. Fong), Rural Industries in China (Tientsin: Chihli Press, 1933). For a more complete presentation of the advocates of this view, see P'eng Tze-yi, ed., Chung-kuo chin-tai shou-kung-yeh shih tzu-liao [Source Materials on the Modern History of Chinese Handicrafts], 4 vols. (Peking: San-lien Book Co., 1957), esp. vols. 2 and 3. It may be noted that Sun Yat-sen, Chiang Kai-shek, and Mao Tse-tung have all supported this view.

proceeded side by side with the observed, relative decline in some rural handicraft industries.[1] The preceeding chapters of this study have clearly demonstrated how such a process has actually unfolded. The substantive findings, when woven together into a trade-equilibrium framework, suggest that the general validity of the "rural disruption" hypothesis cannot be maintained.[2]

What is gained in this perspective apparently also helps to dispel certain ill-feelings, being largely spin-offs of the anti-foreign sentiment, about modern transport development in prewar China. Chiang Kai-shek, echoing an attitude perhaps reflective of that generally held by the politician-intellectual class at the time, charged that the new industrial and commercial system developed in the treaty ports was extended, through railways and steamships, to the Chinese interior, causing "the bankruptcy of [China's] existing handicraft industries and the decline of [China's] agriculture."[3] This assertion, needless to say, can hardly be reconciled with the findings of this study and with the empirical evidence on the long-term transformation of the agrarian economy of prewar China into an increasingly commercialized and spatially integrated agricultural sector which became closely intertwined with a dynamic, expanding

---

[1] Dwight Perkins, using aggregate income estimates, has shown that per capita income in China rose perceptibly, albeit slowly, during the prewar decades (see Perkins, "Growth and Changing Structure of China's Twentieth-Century Economy," in idem, ed., China's Modern Economy in Historical Perspective [Stanford: Stanford Univ. Press, 1975], pp. 115-66). Based on his estimates, agricultural income would have grown more than 18 % between 1914-18 and 1933 while population increased only about 17% during the same period. In Buck's survey, a total of 82% of the rural areas studies also reported some increase in the standard of living over the years prior to the survey (LUSV, p. 459), thus attesting to the general improvement in rural incomes inferred by the findings of this study.

[2] Other scholars have attacked the "rural disruption" hypothesis from other directions. Chi-ming Hou and Kang Chao, for example, have shown that certain handicrafts, notably hand-weaved clothes, remained highly resistant to modern manufacture competition (Hou, Foreign Investment, chapter 7; and Chao, The Development of Cotton Textile Production in China [Cambridge, Mass.: Harvard Univ. Press, 1977], chapter 7. See also Bruce Reynolds, "The Impact of Trade and Foreign Investment on Industrialization: Chinese Textiles, 1875-1931" [Ph.D. dissertation, University of Michigan, 1975]. Potter, on the other hand, argued that handicrafts remained a relatively minor contributor to rural incomes, so that the impact of their decline, even if it had occurred to such an extent as claimed by the advocates of the hypothesis, could not be as devastating as suggested (Jack Potter, Capitalism and the Chinese Peasants [Berkeley: Univ. of California Press, 1968], chapter 8). Dernberger attempted a more thorough attack on the hypothesis by enumerating the possible adverse and beneficial effects of foreign investment and came out with a positive balance (Robert Dernberger, "The Role of the Foreigner in China's Economic Development," in D. Perkins, ed., China's Modern Economy, pp. 19-48).

[3] Quoted in Hou, Foreign Investment, p. 2.

urban-industrial sector by means of modern transportation.[1]

It must, however, be stressed that increased specialization, which was triggered by transport development and the growth of the modern sector, is not without its social costs. In the absence of effective redistribution mechanisms, the gains from trade will be spread unevenly, and certain parties, instead of sharing in the increased wealth, may even be hurt. Thus in the case of early modern China transport development might have enhanced the intersectoral, as well as the interregional, disparity in the distribution of income and caused considerable stress between social classes and among regions.

Because of the small railway network and the lack of intrasystematic transport improvement,[2] the spatial extent of economic benefits emanating from railway construction was rather limited.

---

[1] In contradistinction to the belief of Chiang and many others, there is in fact no lack of evidence that in many rural communities certain handicraft industries, notably hand-weaving, actually benefited from the development of modern transport. The overall strong resistance of hand-weaving to the competition of manufactured imports has been well studied and scholars have offered a variety of explanations for it. (For a summary of these explanations see Reynolds, "The Impact of Trade and Investment," pp. 30-33.) Advances in modern transport, however, has failed to be included as one of the contributive factors in these explanations. The concluding remark from a study on a hand-weaving center in Kuangtung province would perhaps make clear why transport change should be an important factor. It says that "As communication and trade increased the influence of money economy, agricultural crops have become more commercialized than before and through the operation of the price system the peasants have had to earn more money in order to meet their expenses. It is this available labor power and the easy marketing conditions created by the new communication system that has been responsible for the development of hand-weaving." ("Two Hand-Weaving Centers in Southernmost China," in Tawney, ed., Agrarian China, p. 243). Another study in the same volume gave Wei Hsien, Shantung province, as another example in which rural handicrafts prospered on account of the improved condition of communications (see "Rural Auxiliary Occupations in Wei Hsien, Shantung," in ibid, pp. 229-33). These examples made it clear that transport advances benefited many local economies simply because the opportunity for trade had been drastically expanded. To incorporate these cases into the conceptual framework of fig. 14 , two possibilities may be recognized. One involves the outward shift of the production possibility curve AB as the terms of trade for farm products were improved by changes in transport conditions. Another possibility is that transport advances actually improved the allocative efficiency in farm operations so that the production of (Z,F) moved from a suboptimal position inside the curve AB toward a new equilibrium position on AB itself. This latter possibility is indeed likely in view of the findings in chapter IV above. Both possibilities, however, would entail increases in the supply of F and in the consumption of M while the production and consumption of Z either remained unchanged or actually increased.

[2] Largely because of the indifferent attitude of local officials, China's road system rapidly deteriorated in the closing decades of the Ch'ing dynasty. Reports of Customs officials indicated also a continuing decline in the efficiency of local waterways (see Maritime Customs, Report and Return of Trade of Treaty Ports, 1908, part II, vol. 1, pp. 31-32 and Decennial Reports, 1912-1922, pp. 101, 109). Most foreign experts at the time believed that improved roads could have doubled the efficiency of animal-drawn carts (Maritime Customs, Decennial Reports, 1922-31, p. 295). However, there was never any mention of rural communications or

Consequently regions adjacent to railways or the country's few motorways often gained at the expense of other rural areas, which suffered relative decline due to the diversion of trade flows and/or shrinkage in trade hinterlands.[1]

Nevertheless one item in the balance sheet of the costs and benefits of transport improvement weighed heavily on the side of benefits. Technological advance made rail transport the only effective weapon against massive famines which had long been a menace to the lives and welfare of millions in China's history. The fate of those struck by famines, caused by prolonged droughts or deluges, hinged very much on the speed of communications as regards both the transmission of news messages and the delivery of relief supplies. During the great drought famine of 1876-1879 in North China, when the conveyance of relief supplies still relied on human porterage and mules over dirt paths, as much as 13 million people reportedly perished. An analogous famine that broke out in 1920-21, however, resulted in a death toll of less than half a million, thanks to the vital services of the railways.[2] The swift consignment of foodstuff made possible by railways also helped to rein in the tendency for food prices to skyrocket in both the famine-stricken areas and those proferring relief supplies out of meager surpluses. Although it is difficult to estimate the railway-induced savings in the cost of human suffering in years of famine, one thing remains certain. To those who were on the verge of starvation, the benefit of fast relief provision was incalculable.

Constraints on Railway Benefits and Their Impact on Chinese Agricultural Growth

So far this study has been preoccupied with the task of eliciting the beneficial impacts of railway development on China's farm economy during the opening decades of the present century. The

---

local roads in memorials presented by the Ministry of Posts and Communications between 1906 and 1911. In fact, modern motor roads were rarely seen until the 1920s (see Cheng, Influence of Communications, pp. 58-60). But even as late as 1922, it was estimated that only about 500 miles of improved city streets existed and no more than 1,200 miles of dirt road in the country's rural areas were passable year-round (see Tang, Economic Study of Chinese Agriculture, p. 198).

[1] One such example was mentioned in chapter III, which involved the relative decline of the port of Chefoo since the opening of the Ch'ingtao-Chinan railway.

[2] See W. H. Mallory, China: Land of Famine (New York: American Geographical Society, 1926), p. 29. The difficulty of delivering relief materials, especially food, during this devastating famine was greatly aggravated by the huge consumption requirements of the carriers who had to feed on the food designated for the relief.

reason for the apparent neglect of constraints on railway benefits is not that such barriers were negligible; in fact they were not. Rather, the foregoing analyses of empirical evidence were meant to clarify how much rail transport had come to mean for realized agricultural change in the presence of diverse forces that worked against such developments. Assessing the adverse influence of these forces amounts to gauging the additional amount of benefit that railways could have bestowed on the rural economy had such obstacles been removed. Only a few of the more obvious hindrances to the realization of railways' potential beneficial effects are discussed in the following paragraphs, in which attention is given to administrative inefficiency and government's intervention in trade. Both of these provide valuable lessons.

From the beginning the Chinese railway system was plagued with problems of operational and administrative inefficiencies. Before the establishment of unified regulations by the Ministry of Communications in 1915, China's railway system was but a conglomeration of rail-enterprises with disparate administrative and even accounting systems.[1] This disintegration in railway management stemmed only partially from the largely autonomous nature of earlier foreign-owned lines and those built with foreign capital through loan agreements which usually stipulated a distinct administrative structure for the line in concern.[2] A more pervasive cause, however, lay in the gravitation of power to regional, notably provincial, authorities which grew out of the territorial, unspecialized basis of traditional Chinese administration.[3] Not only did this decentralization of political power impede inter-provincial cooperation of private efforts at railway investment,[4] but it also engendered the proliferation of trade-stifling regional tariffs and quotas as regionalism, manifested by the provincial autonomy movement, climaxed in the 1920s and lingered in the 1930s. The lack of administrative unity in the railway system, on the other hand, effectively hindered the development of inter-railway traffic, which was initiated in 1912. This is understandable when freight consignors had to conform to different regulations and to face a bewildering diversity in the classification of commodities and structure of freight tariffs when

---

[1] See Chang Hsin-ch'eng, Chung-kuo hsien-tai chiao-t'ung shih [History of modern Chinese communications] (Shanghai: Commercial Press, 1937), pp. 81-82.

[2] See E-tu Zen Sun, p. 185.

[3] Ibid.; see also Whitney, pp. 84-85.

[4] See E-tu Zen Sun, p. 185.

the goods had to go through different lines.[1]

It was, however, the manner in which the rate structure was determined that carried profound implications for the growth of interregional trade and rural development in this early period of China's railway history. The tapering rate structure, which would encourage long-haul transportation and stimulate rural exports, was not uniformly introduced before 1918.[2] Moreover, most Chinese railways based their rates on average total costs, a policy which grew out of the inflexible attitude of officials to maximizing profits without regard to the elasticity of demand.[3] Since fixed overhead costs invariably accounted for a sizable proportion (often as much as 50%) of total costs in railway operations, the average-cost pricing policy inevitably resulted in artificially high rates. For example, one report estimated that the Peking-Mukden railway could have reduced its rates by 50% and still earned a profit of 10% in the late 1900s.[4] It is impossible to tell whether the financial difficulty encountered by some Chinese railways was itself a product of the high-price policy designed to increase profits. It is, however, revealing to find that the Chinese National Railroads were able to reduce freight rates substantially in the 1930s and actually came out with larger profits than previously.[5] As far as agricultural development is concerned, keeping

---

[1]Ch'en Hui, Chung-kuo t'ieh-lu wen-t'i [China's railway problems] (Peking: San-lien, 1955).

[2]Ibid., pp. 101-102. Tapering rates, meaning that rates charged increased less than proportionally with distances of shipments, conform to marginal cost pricing in the railway industry. This is because (see below) rail operations involve large terminal, fixed costs which can be spread over more miles as shipment distances increased.

[3]See Arthur Rosenbaum, "Railway Enterprise and Economic Development, The Case of the Imperial Railways of North China," Modern China 2 (April 1976):240-41. Theoretically, as long as transport rates are equal to or above marginal cost, financial loss can be avoided in railway operation. The fixed overhead costs, on the other hand, may be recovered from the demand side by charging each traffic the contribution to such costs that it "will bear," i.e. contributions will be taken from each consignor according to his respective demand elasticity (see P. W. Reed, The Economics of Public Enterprise [London: Butterworths, 1973], pp. 144-45). Since agricultural products generally have very low cost-bearing capacity in transportation, their freight rates would be among the lowest even if conscious efforts to stimulate agricultural trade are not undertaken.

[4]See Rosenbaum, pp. 240-241.

[5]Railway freight rates were reduced as much as 80% on the Chinese National Railroads in the early 1930s, partly as a remedial measure to stimulate agricultural growth after long periods of civil war and in the face of a severe depression (see Yu Yen, Tsui-chin-san-nien t'ieh-lu chien-ti yün-chia shu-lüeh [Survey of reduction in railway freight rates during the past three years][Nanking: T'ieh-lu pu, 1935]). From 1932 to 1935-36, however, gross profits of the National Railroads increased from minus (loss) $14.5 million to $45.4 million (see Hou, p. 40, table 10).

freight rates low on the railways could have been the only effective instrument in disseminating trade benefits to isolated rural regions.

It can indeed be said that if the misfortune of prolonged civil wars and rampant warlordism could be avoided, the two decades immediately prior to the outbreak of the Sino-Japanese War would have witnessed considerable progress in the efficiency of the Chinese National Railways. In chapter II above we have already mentioned the havoc wrought by successive military interference on Chinese railways during the late 1920s. It suffices to add here that as a result of war-related damages, the operating efficiency of the National Railways remained at only 10.4% as late as 1933, meaning that the average performance rating of the railways in freight shipment was merely one-tenth of the maximum efficiency under optimal operating conditions.[1]

If political strife and internal warfare represented an inexorable course of events that engulfed the country, there were lost opportunities that could have mitigated the plight of China's railway system. The Chinese government, by siphoning off a substantial portion of the railways' earnings, displayed little initiative in aggressive marketing and service innovations, much less a commitment to capital improvement. The gross inadequacy of storage facilities, for example, was both a consequence of the underinvestment in auxiliary structures and an apathy towards catering to customer needs.[2] The net effect, needless to say, was to discourage commercial freight shipments via railways, which were already adversely affected by the deterioration in operational efficiency due to the lack of maintenance on rolling stock.[3]

Ironically, one important source of capital for necessary maintenance works could have come from the railway system itself. Among the expenses of the Chinese railways, the amounts paid for the military force employed for guarding the lines represented a very important item.[4] Most waste, however, was a direct result of mismanagement. According to a foreign commission's report, as much as 40%

---

[1] See Ch'en Hui, p. 108.

[2] An early illustration is provided by the case of the Peking-Mukden (North China Imperial) Railway (see Rosenbaum, pp. 237-38). For another example in the 1930s, see the discussion in Sung Chiang Mi-shih t'iao-ch'a [A study of the Sung Chiang rice market] (Shanghai: The Institute of Social and Economic Research, 1936), p. 25. See also the general appraisal of the situation in Shen Tsou-t'ing, T'ieh-lu wen-ti tu-lun chi [Essays on railway problems] (Shanghai: Commercial Press, n.d.), pp. 112-114.

[3] See de Fellner, pp. 116-17.

[4] Ibid., p. 119.

of the operating expenses of the Chinese National Railways could be categorized as unnecessary or waste.[1] The sources of such waste included the underemployment of railway staff, mishandling of fuel by unskilled employees, and the abuse of rolling stock.[2] A rationalization of the distribution of capital outlays also could have improved operating efficiency and contributed greatly to service improvements.[3]

In broader perspectives, however, the difficulties in which the Chinese National Railways were entrapped were but one illustration of the general inability of the Chinese government to provide for the necessary conditions favourable to modern economic growth. Given the interdependence between the urban-industrial and rural sectors, the government's failure to promote industrialization at an early date must also have stripped agriculture of the opportunity to embark on early modernization. As Perkins has argued persuasively, even if the Ch'ing government could not have contributed directly to industrialization because it was financially pinched, it could still have promoted institutions that enhanced private capital formation and launched efforts to promote technological change.[4] In both cases, however, the government failed to assume an active role.[5] The primitive organization and slow development of the banking institution in early twentieth-century China in particular proved to be a strong hindrance to commercial development, with the difficulty of financing private railway construction standing as one particular example.[6] The Republican government, established in 1912, inherited the economic problems of the late Ch'ing, its incompleted reforms,

---

[1] See Ch'en Hui, pp. 107-8.

[2] Ibid.; and de Fellner, p. 120.

[3] When compared with contemporary American and Japanese railways, the Chinese National Railways were characterized by excessively high disbursements towards covering such indirect cost items as general administration and wages of crew and guards. Expenditure on items which are directly related to service improvement and operational efficiency was, however, highly inadequate (see Ch'en Hui, pp. 105-6; and Cheng Ching-lieh, "Wo-kuo t'ieh-lu ying-yeh yung-k'uan chi k'ai-kuang chi chi chen-li chi kuang-chien [A survey of the operating expenses of Chinese railways and an overview of its improvement]," Chiao t'ung ch'a-chi 1 (1933), p. 80).

[4] Dwight H. Perkins, "Government as an Obstacle to Industrialization: The Case of Nineteenth-Century China," Journal of Economic History 27 (Dec. 1967):487-88.

[5] Ibid.

[6] See E-tu Zen Sun, p. 187, for some examples of the futility to raise capital for private-motivated railway schemes in a situation where financial institutions were underdeveloped.

and its lack of financial stability. Little change was visible in
the passive role of the government in promoting industrial development. Indeed, until the National Government finally assumed nationwide responsibility in 1928, it was turmoil, more than anything else,
that the politically impotent Republican government had contributed
to the economic sphere.

It was not, however, what the governing authority in China had
failed to do to the urban-industrial sector that had the most profound, restrictive influence on agricultural growth and the benefits
of railway development. The successful exploitation of railways as
an instrument for agricultural growth is heavily conditioned upon
the development of supporting transport infrastructures and the
elimination of existing barriers to interregional trade. In both
cases the government failed. As mentioned earlier, China's traditional highway system had deteriorated to a deplorable state in late
Ch'ing times, largely as a result of the lack of official attention.
The situation was not improved since the establishment of the Republican government. Before 1920, when North China was stricken with
the most severe famine since 1879, practically nothing had been done
towards building roads suitable for motor trucks or even for heavy
cart traffic.[1] Back of the expansion of railways therefore lay a
pervasive underdevelopment of auxiliary transport infrastructures.
Consequently the spread of economic incentives emanating from railway development was severely restricted.

The general improvement in local transport conditions, if it
had indeed occurred on a large scale, would only be able to stimulate the growth of interregional trade marginally with as cumbersome
a system of transit duties as the likin of prewar China in existence.
The likin system, first instituted in 1853 as a local, temporary
financial measure, soon became deep-seated as an important source of
revenue to provincial governments. From the beginning, likin was as
much a nuisance to traders as a deterrent to trading activities.
Because of the absence of uniform regulations and centralized control, duty rates were as diverse as the commodities subject to the
levy and the ways of actually exacting the taxes many times over.[2]
Abuses were rampant, as duty assessment was subject to arbitrary

---

[1] Mallory, p. 32. It may be noted that road maintenance was often socially desirable projects in view of the relatively low costs (only about C$100 per mile per year in North China in the 1920s [see Cheng Ming-ju, p. 62]) and potentially high returns to farmers and merchants making use of the roads.

[2] See D. K. Lieu, China's Industries and Finance (Shanghai: Chinese Government Bureau of Economic Information, 1927), pp. 129-34.

arrangement by individual collecting officials.[1] Compounding the vexations to merchants was the multitude of collectorates along the same trade route, which inevitably resulted in prohibitive, accumulative rates of tariff. An estimate by Lieu put the number of likin collectorates in 1927 at 790, and sub-collectorates at 5,300.[2] From Shanghai to Chinchiang, a distance of less than 200 miles, goods conveyed by non-railway modes of transport in the mid-1900s had to pass through as much as seven likin barriers, paying a real rate of 35% or more of their original prices.[3] For longer distance shipments, it was not uncommon to find cumulative tax levies close to 30% of the gross value of the commodities even after the general reduction of rates since the establishment of the Republic.[4]

Railways, of course, were not exempt from the interference of the likin, but the coming of the railways did help promote limited reforms in the system, at least along railway lines. Goods on board freight trains from Shanghai to Nanking, for example, needed pay only a total tax of 13.5% on a uniform rate of 1.5% for each of the nine sections along the railway, instead of the much higher 35% or more that would have been collected along a river route.[5] As another example, goods transported from Hankow to Peking via the Peking-Hankow railway would have to pay likin only twice, at a rate of 2.5% each, thus avoiding the insurmountable barrier set by the several hundred likin collectorates along the trade route.[6] Nevertheless rail transport suffered greatly during the height of the civil war period when railway revenue was eagerly requisitioned by warlords and the government for military finance. The willful exploitation of the railways as a source of revenue resulted in large increases in both the tax burden and the number of collectorates along many rail routes.[7] Needless to say, the net

---

[1] See Morse, pp. 104-7; see also Wagel, pp. 380-85.

[2] Lieu, p. 130.

[3] Ibid, p. 135.

[4] For examples see CKNY 2:284-86.

[5] Lieu, p. 136.

[6] Ibid., pp. 147-48.

[7] At one time, as much as 82 collectorates were staged along the Peking-Suiyuan line for the purpose of military finances. Tax rates on the Ch'ingtao Chinan line, as another example, were suddenly increased ten times for similar purposes (see CKNY 2:283).

effect of such actions was to greatly deter the long-distance trading of commodities, but in any case interregional trade was already smothered by the frequent, arbitrary imposition of embargoes and quotas on provincial import-exports by warlords.[1]

Apparently, the imposition and continued existence of likin, as well as any other form of domestic trade barrier, hindered both interregional trade and the internal integration of the economy. It suppressed the economic opportunity of farmers by distorting the incentives for trade and thereby depressed agricultural output. The likin, therefore, tended to nullify the benefits that railway and general transport development would have brought to the rural sector. It is indeed lamentable that likin, frequently in new disguise, lingered long after its official abolition in 1931.[2] Apart from political instability, the likin was perhaps the most important cause of divergence of potentiality and actuality in railways' contribution to agricultural trade and rural economic progress.

---

[1] CKNY 3:144-148.

[2] CKNY 3:133-37. See also Chang Chih-i, Fukien-sheng shih-liang chi yün-hsiao, pp. 92-96.

CHAPTER VII

CONCLUSIONS

In an article that summarizes the comparative success of railway operation in India, China, and Japan, Abramovitz concluded that railways in prewar China had largely failed in their role as an instrument in the process of national economic growth.[1] Few, indeed, would dispute with Abramovitz on the observation that China's railways had minimal impacts on the country's development of technological and business know-how, or even the expansion and internal integration of the country's financial institutions. Nor had Chinese railways contributed much to the process of industrialization through increased demands for manufactured products used in railway construction. Yet these aspects, as Abramovitz emphasized, were exactly what characterized the important role of railways in the economic growth of many countries.

Nonetheless Abramovitz's focus may have been misplaced. If China's railways had failed in many other aspects, it was certainly in agriculture that they had scored the most conspicuous success. In a country in which agricultural production accounted for nearly two-thirds of national income, the contribution of railways to Chinese economic growth could be considerable simply on account of the beneficial impact on agriculture. Through careful examinations of empirical evidence, this study has been successful in identifying some of the most important aspects of the beneficial impact that railway development had on the agrarian economy of prewar China.

Perhaps the most visible influence of the railway on China's peasant economy was the imprint it left on the direction and magnitude of the interregional flows of agricultural goods. In areas where the railway came into existence, it often evoked significant shifts of trade toward a railbound course. However, the more profound effect emanated not from the mere diversion of trade, but from the increased production and marketing of farm products in areas taking advantage of new trade opportunities opened up by the railways. As a result, the coming of the railway added much complexity

---

[1] Abramovitz, "Railroads and Economic Development."

to the originally river-dominated pattern of domestic agricultural commerce. By stimulating commercial cropping and regional specialization in production, the railway also introduced much diversity to the composition of agricultural trade which was once characterized by a few luxury commodities. By the mid-1930s, probably close to two-thirds of China's long-distance agricultural trade in commercial crops (by volume) were rail-borne. Since rice was perhaps the only subsistence crop that also entered significantly into long-distance trade with a dominant carrier other than railway, rail transport played a vital role in the observed expansion of long-distance commerce in prewar China.

The extent to which railway development affected agricultural production constituted a major focus of investigation in this study. A series of hypotheses was introduced and empirically examined by means of econometric and statistical techniques. The most important conclusions derived from these analyses include (1) the accessibility (economic distance) of a farm to the regional market of sale had an important effect on the level of farm production. The closer a farm was (in terms of economic distance) to the market, the more likely was the farm to have a higher level of "normal" output. The reason is simply that better accessibility, ceteris paribus, means higher farm prices, which stimulated the added employment of production resources as well as a more efficient utilization of these resources in production. (2) Railway construction in prewar China appreciably improved the market accessibility of many rural areas. This, in turn, resulted in increased farm production in the affected areas. (3) Given the fact that there was no observable difference in land productivity (in the pre-railway period), between areas affected and those not affected by railway building, the net effect of railway construction was to raise the average level of farm production and land productivity in the areas directly affected by railways, so that both average farm output and land productivity in the railway areas were significantly higher than in the non-railway areas. (4) Through its effect on land productivity, railway construction exerted upward pressures on the value of agricultural land. In the short run, railway building was usually attended by a discrete jump in local farmland values. In the longer run, however, both the increased accessibility of the affected locality and the improvement in local farm productivity attracted an increasing inflow of migrants. Because the opportunity for alternative non-agricultural employment was usually scarce, the influx of migrants was translated into rising population pressure and a faster growth in land values when compared with land value movement in the pre-railway period.

An important factor underlying the rail-induced changes in agricultural production was the stimulative effect of railway construction on commercial cropping and the enhanced participation of peasants in the market economy. Study of sample areas revealed that the spatial affinity of greater specialization in commercial cropping to railway routes was often strong. Empirical evidence also abounds in support of the expected increase in the percentage of local cultivated areas devoted to commercial crops following railway construction. These findings fall in line with the identified strong empirical relationship between increases in market accessibility and the size of the "marketable surplus" of farms. In light of the conclusions arrived at earlier, railway development in prewar China carried not only important implications for agricultural development, but also for urban and industrial growth because of the effect on agricultural surplus, i.e., that part of the agricultural output in excess of the consumption demand of the farm sector.

Two important qualifications on the foregoing statements on the economic effects of railways must be emphasized, however. First, the impact of railways on agriculture had a distinctive regional dimension. The extensive network of natural and artificial waterways and the high density of human settlements in many parts of South China (the lower Yangtze territories in particular) rendered traditional water transport a formidable rival to rail transport. In these areas, railways generally gained importance only in the carriage of higher-valued farm products, mainly by virtue of the higher speed and the lower risk of loss and spoilage of cargo en route. In most of North China, where navigable waterways were scarce and the average haul between shipment points and market destinations usually long, rail transport had a leverage not only in terms of speed and safety, but also in terms of unit charges (freight rates). In fact, the competitiveness of river shipping in South China very likely discouraged the expansion of the regional railway network and greatly limited the observed aggregate impact of railways on South China's agriculture. The estimated incremental gain in agricultural output in South China due to rail construction, as derived in chapter IV of this study, was in fact less than 2%, as compared with North China's 13%.

Another qualification concerning railway effects relates to exogenous factors. One driving force of the spread of commercial cropping along railway lines was the consistent endeavor on the part of foreign enterpreneurs to promote the commercial cultivation of imported crop species at accessible railway sites. The most notable example was of course provided by the British American Tobacco

Company and the Nanyang Brothers Tobacco Company. Both the private capitalists' and the government's propensity to encourage commercial cropping along rail transport routes reinforced the tendency of improved foreign crop species to diffuse along railway lines. This in turn contributed to increases in land productivity and improved opportunity for obtaining institutional credit in railway areas.

There were, however, external forces that operated against the realization of potential benefits flowing from railway construction. Political disorder and civil warfare were frequent menaces to the efficient operation and planned development of the railway system throughout the prewar decades under consideration. But equally obstructive to rail transport efficiency improvement were the prevalence of mismanagement and apathy toward capital improvement and aggressive marketing on the part of railway officials. The perversity of the government's role in fostering the spread of railway benefits in fact appeared most distinctly in the persistence of the likin or transit duties which, because of their ubiquity and abuses, had lingering, ruinous impacts on agricultural development and the growth of interregional trade. The net effect of the existence of the likin, indeed, was to gravely undermine the beneficial influence emanating from railway development.

Nonetheless, the realization of the presence of adverse influences only enhances the appreciation of railways' potential contribution to the prewar Chinese economy. A main objective of this study has been to demonstrate how much railways had come to mean for realized agricultural improvement even in the presence of multiple constraints. Findings which confirm positive impacts of railways on agricultural productivity can easily be woven together in a conceptual framework which shows how the rural sector might have gained economic progress under the trade mechanism. In the light of empirical evidence, this conceptual framework poses direct challenges to the often-held notion of rural impoverishment in the face of modern sector development and increased direct foreign investment in prewar China. Rural decline, if it had indeed occurred, was more likely a result of too little, instead of too much, investment in modern transport infrastructures in which the foreigner played a vital role.

APPENDICES

APPENDIX A

ESTIMATION OF EXPECTED AVERAGE FARM OUTPUT

Total expected farm output per average farm in a locality was estimated by the following equation:

$$Y_k^* = \sum_i x_{ik} q_{ik} p_{ik} + \sum_j h_{jk} u_j p_{jk}$$

The definitions and sources of these variables are as follows:

x_{ik}: The average cropped area devoted to crop i per average farm in locality k, was calculated from Buck, LUSV, pp. 174-79, for each of the 28 most commonly cultivated crops;[1]

q_{ik}: Most frequent yield[2] per hectare of crop i at locality k, was obtained from Buck, LUSV, pp. 209-10 for 20 of the 28 crops. For the remaining 8 crops, many of which were observed to be cultivated in only a few localities, provincial or even national average yields were used;[3]

h_{jk}: The average number of the j-th type of livestock per farm in locality k, was obtained from Buck, LUSV, pp. 122-23 for six major types of livestock (i.e., j = 6);[4]

---

[1] The 28 crops are barley, buckwheat, corn, kaoliang, millet, proso-millet, rice, glutinous rice, wheat, broad beans, field peas, soybeans, black beans, green beans, peanuts, rape-seed, Irish potatoes, sweet potatoes, cotton, hemp, sesame, fruits (all kind), vegetables (all kind), tea, tobacco, sugar cane, grains interplanted with legumes (corn and soybeans, kaoliang and soybeans, wheat and broad beans, others), and mulberry (silkworm cocoons). All together these crops accounted for more than 92% of the total cropped area in the country.

[2] Most frequent yield of crops found on 20 per cent or more of the farms, as obtained from over a period of 10 years of observation. Assuming that yields in the 10 year sample were not highly skewed in distribution, the most frequent yield (mode of distribution) may be taken as the average yield (mean) over the ten-year period.

[3] The following sources are employed: (a) Fruits and vegetables: Ministry of Industry, Chung-kuo ching-chi nien-chien [Chinese economic yearbook] (Nanking: Government Printing Office, 1934); (b) Tea: Chu Mei-yu, Chung-kuo ch'a-yeh; (c) Silkworm cocoons: Shen Tsung-han, Chung-kuo nung-yeh tzu-yuan, 2:64; (d) Hemp, sesame and sugar cane: Liu and Yeh, Economy of Chinese Mainland, p. 300. Seed cotton was converted to lint cotton equivalents by assuming a yield ratio of 30: 100 between lint and seed cotton. Crops interplanted were treated as having equal shares in the interplanting crop-hectare areas.

[4] The six kinds of livestock were hog, sheep, goat, geese, chicken and duck. Only adult animals were counted.

u_j: The utilization rate of the j-th type of livestock (including the production rate of non-meat animal products),[1] was derived from Liu and Yeh, Economy of Chinese Mainland, pp. 311-14;

P_{ik}, P_{jk}: Prices of the i-th crop and the j-th type of livestock, respectively, were expressed in kg. grain-equivalents (i.e., in kg. wheat and rice equivalents in the Wheat and Rice Regions respectively). Provincial averages of farm prices in 1933 were used and obtained from Liu and Yeh, Economy of Chinese Mainland, pp. 319-59.[2]

---

[1] The utilization rate of a certain kind of livestock was derived by Liu and Yeh based on the average slaughtering rate and estimated rate of sale of the bodies of animals died of sickness and old age. In the present calculation the utilization rate also included the production rate of non-meat animal products for chicken (eggs), sheep (wool), and goat (hairs). The respective production rates were obtained from Liu and Yeh, Economy of Chinese Mainland, pp. 309-14.

[2] The use of provincial average prices instead of local prices helped minimize the risk of identifying a spurious relationship between local output and accessibility. To the extent that peasants substituted cash crops for subsistence crops more readily in more accessible areas, the use of prevailing local prices, which would be positively related to the proximity of the farm to the market of sale and more so with more profitable crops, would have built in the covariational relationship between farm output and accessibility which the analyses in chapter IV seek to identify.

APPENDIX B

ESTIMATION OF THE ACCESSIBILITY INDEX

The accessibility index is defined as the total cost of transporting a ton of farm products from the farm to the local market and from the local market to the regional market, plus applicable loading, unloading and transhipment charges.

The local market is assumed to be the hsien capital. The cost of transport from the farm to the hsien capital represents the least total cost of shipment among various common modes of local transport, provided the average trip distances reported for the various modes were reasonably close. If such trip lengths varied considerably among modes, the cost is a weighted average of the costs of the trips by different modes, the weights being chosen in inverse proportion to the length of the trip. Basic data on local transport rates and marketing distances pertaining to various common modes of transport were from LUSV, pp. 344, 346.

Transport cost from local to regional markets was estimated as follows. First regional markets were identified according to the following criteria: (a) A regional market must be a regional, preferably provincial, center of collection, distribution, or consumption of at least one major interregionally traded agricultural commodity (see fig. 5 for basic data and sources); (b) No more than two centers would be chosen in any one province. Specifically, only the two most important centers would be selected in cases where more than two existed. (c) If no ordinal ranking of trade centers was possible in any one province in the Rice Region, the largest rice market in the province would be selected (assuming one existed). No similar criterion applies to the Wheat Region, since neither rice, nor any other foodcrop, was as heavily traded in North China as was rice in South China. Normally, major transport and population centers coincided with regional markets in the Wheat Region.

The relatively simple pattern of interregional trade flows in prewar China actually offered little difficulty for the identification of the most important regional markets. The result of the identification procedure yielded a total of seven markets in the Wheat Region and thirteen in the Rice Region.[1]

---

[1]Wheat Region centers are: Paot'ou (Suiyuan), Tientsin (Hopei), Chengchou

The shipment distances along existing routes from the local to the regional markets were directly estimated from atlases.[1] Water transport rates along principal routes were gathered from various sources and the most common rate on a particular route was determined and applied to all water-borne freight on that route. Rail freight rates were obtained from Ministry of Communications, Republic of China, Chiao-t'ung shih, Lu-cheng p'ien [Section on railways and motor roads, history of communications in China] (Nanking: 1930). Transport rates pertaining to all other modes of transport were from Buck, LUSV, p. 347.

Information on loading and unloading charges of river transport are difficult to obtain, but where they are available, such charges usually amounted to 7% - 13% of the total transport costs, depending on the distance of the trip.[2] For consistency in estimation, a 7-8% charge was applied to trips over 300 miles, a 9-11% to trips between 100 and 300 miles, and a 12-13% charge to trips under 100 miles, with the exact charge in each case determined by the topographic conditions along the route and the direction (upstream or downstream) of the trip. Railway terminal charges were obtained from the same source as the freight rates, which varied slightly among lines but mostly amounted to about C$0.10 per ton.

(Honan), Ch'ingtao, Chinan (Shantung), T'aiyüan (Shanhsi), and Pangpu (Anhuei). Rice Region centers are: Shanghai, Wuhsi (Chiangsu), Hangchou, Ningpo (Chechiang), Wuhu (Anhuei), Wuhan (Hupei), Nanch'ang (Chianghsi), Fuchou (Fuchien), Canton (Kuangtung), Ch'ungch'ing (Ssuch'uan), Wuchou (Kuanghsi), K'unming (Yunnan), and Kueiyang (Kueichou).

[1]Chang Ch'i-yün, Chung-hua min-kuo ti-t'u chih [Atlas of the Republic of China], 5 vols. (Taipei: National War College, 1963); and Ting Wen-chiang et al., eds., Chung-hua min-kuo hsin ti-t'u [A new atlas of the Republic of China], 2d ed. (Shanghai: Shen Pao Press, 1934).

[2]See, for example, Chang P'ei-kang, Kuanghsi shih-liang wen-t'i, pp. 120-23.

APPENDIX C

ESTIMATION OF THE VALUE OF MARGINAL PRODUCT PER
MAN-WORKDAY AND DAILY LABOR COST

The value of marginal product (VMP) per man-workday is defined as the addition to farm output, valued at the prevailing output price, resulting from the use of one more man-workday in the production process. The VMP per man-workday was estimated from the VMP per man-equivalent derived from the production function analysis summarized in table 10 (chapter IV). VMP per man-equivalent was obtained by multiplying output price (wheat and rice prices in the Wheat and Rice Regions respectively) by the marginal product (MP) of labor (in man-equivalents). The marginal product of labor was in turn the product of the output elasticity with respect to labor (from equations 2'a and 2'd of table 10) and the average product of labor (i.e., output/labor). The VMP per man-equivalent in the Wheat and Rice Regions are given in table 21, column (3).

In order to translate VMP per man-equivalent into VMP per man-workday, the former has to be divided by a suitable figure which represents the average number of days in a year in which active farm works were carried out. Since the average number of non-idle days for farm laborers in prewar China was reported to be around 300,[1] and the average duration of the peak farming seasons was four months, or about 120 days,[2] a simple average of these two figures gave 210 days as the average number of days of active farmworks.[3] From this

---

[1] According to Buck (LUSV, p. 307), each male adult unit had two idle months in an average year in both the Wheat and Rice Regions. This gave ten months, or approximately 300 days, as the average length of a work-year.

[2] In the Wheat Region, the months of April, May, August and September each had less than 2% of the total idle labor time in a year and can be considered as the peak farming seasons (two two-month periods) (see LUSV, p. 328). In the Rice Region, the first peak season occurred from early April and last till the end of May. The second peak season usually started in early September and last through October when harvesting and late-planting for a second crop were carried out (ibid.). In general, therefore, four months or 120 days may be designated as the length of the peak farming season.

[3] The simple average is justified on the ground that slightly more than half of the total farm output was realized during the peak farming seasons (see below).

TABLE 21

TESTING ALLOCATIVE EFFICIENCY OF LABOR IN CHINESE AGRICULTURE

Region (1)	MP (2)	VMP (3)	Survey year Food Price Index (1933 = 100) (4)	Cost of year Labor		Cost of labor (Deflated)		Estimated Opportunity Cost (Day Labor) (9)	VMP/ man-day (10)	(9)-(10) (11)	t (12)
				Survey (5)	Deflated (6)	Daily (7)	Monthly (8)				
Wheat	421.67	37.95	135.60	75.22	55.47	0.324	6.431	0.189	0.181	0.008	0.65*
Rice	366.14	25.63	166.67	87.47	52.48	0.282	5.124	0.146	0.122	0.024	2.67**

SOURCES: Col. (2), the marginal product of labor, is in kg. grain-equivalent and is derived from multiplying the average product of labor by the output elasticity with respect to labor as estimated in table 10, equations (2"a) and (2"d); col. (3), the marginal value product of labor, is the product of col. (2) and the 1933-34 prices of wheat and rice in the Wheat and Rice Region respectively; col. (4), the general food price indices pertaining to 1929-31, were computed from Shanghai Economic Research Office of the Chinese Institute of Science, Shanghai chieh-fang ch'ien-hou wu-chia tzu-liao hui-p'ien [A Compilation of Price Materials for Pre- and Post-Liberation Shanghai] (Shanghai: People's Press, 1958), pp. 175, 126 and 213; col. (5) is from Buck, LUSV, p. 328; col. (6) = col. (5)/col. (4); col. (7) and col. (8) are from Buck, LUSV, p. 328 and have been deflated by the index in col. (4); col. (9) = (7) x [((7) x 120 + (8) x 4)/ (5) x 2]; col. (10) = col. (3)/120.

Note. - Unless otherwise stated, all labor costs are in C$.

*Null hypothesis of no difference between cols. (9) and (10) accepted at 5 % level of a two-tailed test.

**Null hypothesis accepted at 0.1% level of a two-tailed test.

the VMP per man-workday was estimated to be 0.181 and 0.122 (in kg. wheat and rice equivalents) in the Wheat and Rice Regions respectively (see column [10], table 21).

For approximate measurement of the opportunity cost of farm labor, the wage data given in LUSV, pp. 328-29 were utilized. By deflating these cost figures, which pertained chiefly to the years 1929-31, to the general price level in 1933 (the year on which the present analysis is based), it becomes meaningful to compare them with the estimated VMPs of labor. In column (6) of table 21 were displayed the deflated average annual cost of labor in the Wheat and Rice Regions. Comparing these figures with the corresponding estimates of VMP per man-equivalents given in column (3), it appeared that the resulting differences were too large to justify the existence of allocative efficiency in the use of Chinese farm labor.

However, it may be argued that farm outlay on year labor is not an appropriate measure of the opportunity cost of labor in the present case. Laborers hired on a long-term yearly contract were often treated very much like part of the farm family, receiving not only wage but also shelter, food, and other daily necessities from their employers. In return, they were often asked to help out in subsidiary non-farm works during slack seasons. As the VMP of labor estimated in this study refers to farm works only, annual labor cost tended to overstate the opportunity cost of farm labor hired for farm works alone.

For this reason seasonal labor cost may provide a better measure of the opportunity cost of labor than annual labor cost. Since seasonal labor was confined to peak farming seasons only, the daily cost of hired farm labor would have to be weighted by the proportion of the total agricultural work performed in the peak seasons to arrive at an approximation of the true year-round opportunity cost figure. This of course assumed the opportunity cost of farm labor in the slack seasons to be zero, which may not be unreasonable considering the general absence of alternative non-agricultural employment opportunities for farm labor in the slack seasons.

As pertinent data on seasonal farm output cannot be obtained, an estimation of the proportion of farm work done during the peak seasons was based instead on the assumption that actual monetary return to farm labor on a daily basis was indicative of, or at least proportional to, the amount of farm work actually performed. The average seasonal outlay on day and month labor as a proportion of the outlay on year labor was therefore taken as the proportion of farm work accomplished in the peak seasons. As a result, 58.3% of total farm output was estimated to be accomplished in the peak seasons in the Wheat Region, whereas in the Rice Region the estimate was 51.8%. These figures were then used as weights to derive the daily opportunity cost of labor, as given in column (9), table 21.

Comparing the estimated daily opportunity cost of farm labor with the VMP per man-workday figures, it was found that no significant difference could be detected between the two in both the Wheat and Rice Regions (see columns [11]-[12], table 21). At least at the regional level, therefore, farm labor was found to be fairly efficiently allocated in farm production in prewar China.[1]

---

[1]This result is in general agreement with that obtained by Myers (Ramon Myers, "Resource Allocation in Traditional Agriculture: Republican China, 1937-40," Journal of Political Economy 79 [July 1971]:25-36.

APPENDIX D

THE ENDOGENEITY VS. EXOGENEITY OF RAILWAYS AS
A FACTOR IN AGRICULTURAL CHANGE

The existence of systematic differences in agricultural productivity in areas later affected by the construction of railways and those that were not so affected is immaterial to a study which focuses on the actual changes in agricultural productivity brought about by railways in affected areas. However, any inference about such productivity changes which is based on cross-section analysis may be badly flawed if the spatial configuration of railway investment was dictated by geographical disparities in agricultural productivity instead of being the generator of such productivity diferences. Although the present study has not relied on cross-section data as the sole source of empirical evidence, the existence or non-existence of *ex ante* (before-railway) disparities in the agricultural productivity of railway and non-railway areas[1] remains a question crucial to a better understanding of the magnitude of railway-induced agricultural changes.

Before appealing to empirical data, a few observations on historical conditions may be worth mentioning. First, the very design of the railway network in early modern China was influenced as much by political as by economic considerations. This was true, for instance, of the Chinese railway building in Inner Mongolia and Manchuria, which had the purpose of containing foreign influence in these frontier regions by way of strengthening the economic and political ties between these segments of the periphery and the center.[2] Thus both the Suiyuan-Peking and Peking-Mukden railways traversed territories which were economically undeveloped at the time of the railways' construction. The same can also be said of the Yunnan-Indochina railway built by the French in the southwest. The ultimate goal of the French in such an endeavor was not so much the economic development of Yunnan province as the acquisition of an

---

[1] See the definition on p. 90 in the text.

[2] For discussions of the Inner Mongolia case, see O. Lattimore, "China and the Barbarians," in *Empire in the East*, ed. Joseph Barnes (New York: Doubleday, 1934), p. 28. For the Manchuria case with respect to the Chinese Eastern Railway see Kent, pp. 45-46.

assured access to the resource-rich markets of Ssuch'uan and the middle Yangtze valley through subsequent northward extension of the railway.[1]

Second, many of the early lines, particularly those which originated with foreign demands for railway rights within respective foreign "spheres of influence," were so designed as to further development of trade between established commercial centers.[2] The perceived hinterland of importance to foreign investors in many early railway schemes was mainly extraregional.[3] To a considerable extent, therefore, existing local economic potentials did not become explicit determinants in the routing of trunk lines. Oftentimes, of course, express concern with intraregional productivity in agriculture was unnecessary because of the relative homogeneity of conditions within the region traversed by the proposed line. In these cases productivity variations between near-railway and off-railway areas can also be assumed to be statistically insignificant.

Finally, several major railway lines actually cross regions that were agriculturally disadvantaged because of poor physiographic or soil conditions. For example, both the Peking-Suiyuan and the French Yunnan-Indochina lines pass through largely mountainous territories. This is also true of the Chechiang-Chianghsi railway, though perhaps to a lesser extent. The Lung-Hai line, on the other hand, runs along the southern flood-plain of the Yellow River and traverses zones notorious for poor fertility.[4] The same soil type also characterized the narrow strip of land in western Hopei, on which runs the Peking-Hankow railway.[5] In comparative terms, therefore, communities along the Hopei section of the Peking-Hankow railway were likely to be less productive agriculturally, ceteris paribus, than their counterparts lying on the fertile plain region to the East.

These observations notwithstanding, the existence or non-existence of *ex ante* differences in agricultural productivity in

---

[1] See E-tu Zen Sun, pp. 181-82.

[2] See ibid., p. 180.

[3] For example, the German-built Ch'ingtao-Chinan line was originally built as a feeder to the Peking-Hankow railway so that "Kiaochou (Ch'ingtao) could be made the terminus of a line to Peking and North China" and that "the Power which possesses Kiaochou will control the coal supply in northern Chinese waters." (Quoted of Baron von Richthofen in Kent, p. 147.)

[4] See Buck, Land Utilization, pp. 136-38, and George B. Cressey, Land of the 500 Millions (New York: McGraw Hill, 1955), p. 112.

[5] See the soil types map in Cressey, p. 111.

railway and non-railway areas remains an empirical question. The difficulty of tackling this question stems mainly from the absence of comprehensive data on the direct indicators of local agricultural productivity in the pre-railway period. However, proxies for these indicators can be constructed from what is available. Barring periods of social upheavals, rural population density usually closely reflects the supporting capacity of local land resources. But even population density figures were scarce. More complete data in this respect were compiled from provincial gazetteers of only three provinces which had relatively large railway trackage per square kilometer by the early 1930s. In each of these provinces, namely Hopei, Shantung, and Chechiang, average population density by county was grouped into railway and non-railway categories in accordance with the definitions given in p. 90 in the text. A t-test was applied to the two samples in each of the three provinces, and the results are given in table 22. It was found that the railway <u>hsien</u> in Shantung appeared to be more densely settled in general than non-rail <u>hsien</u> but the fact that the Ch'ingtao-Chinan railway was opened in 1904, or two years before the population registration reported in the gazetteer, has to be taken into account. Neither in Hopei nor in Chechiang was any significant difference detected between the rail and non-rail <u>hsien</u> as far as population density is concerned.

TABLE 22

TESTING POPULATION DENSITY DIFFERENCES
IN RAILWAY AND NON-RAILWAY HSIEN

	Railway Hsien		Non-Railway Hsien		
	Sample Size	Mean Density (person/mou)	Sample Size	Mean Density (person/mou)	t-Ratio[#]
Hopei	42	0.364	87	0.356	0.952
Shantung	23	0.244	87	0.125	2.101*
Chekiang	22	0.385	53	0.407	1.011

SOURCES: Hopei: <u>Chi-fu t'ung chih</u> (1884 edition), vols. 94-96; Shantung: <u>Shantung t'ung chih</u> (1910 edition), vols. 78-82; Chechiang: <u>Chechiang t'ung-chih</u> (1899 edition).

[#]$t = (\bar{X}_1 - \bar{X}_2)/[(s_1^2/n_1 - 1) + (s_2^2/n_2 - 1)]^{\frac{1}{2}}$, where $\bar{X}$, s, and n denote sample mean, standard deviation, and sample size, respectively.

*Significant difference between the two sample means accepted at the 0.05 level of confidence.

The population density test, however, has some major difficulties. First, as Ping-ti Ho observed, the population censuses in late Ch'ing were often regionally inconsistent and produced confusing results.[1] Second, the recorded figures did not distinguish between market town and farm population. Finally, there is the prevalent problem of land under-registration in pre-Republican China. In particular, the Ch'ing land returns were figures on fiscal acreage, not actual cultivated acreage.[2] The weight of these factors combined might well render the population density figures thus derived highly vulnerable to large margins of error.

Apart from population density, the only other plausible basis on which agricultural productivity may be inferred is land tax returns. Land tax was the single most important source of government revenue in Imperial agrarian China, and the Ch'ing Emperors had kept extensive records on it. Unfortunately, though the levy of land tax was in proportion to the rental value of the land, the rental value had been settled some 300 years ago by the Manchu Emperor under a system known as permanent settlement.[3] Under this system land tax rates (quotas) were fixed once and for all. The tax quotas, at least in theory, remained unchanged in spite of rises in land rental values over time. In practice, however, local authorities were able to bleed additional payments from the peasants through a variety of surcharges,[4] especially during the last decades of the Ch'ing dynasty when the government was increasingly pinched with financial woes. Though the level of official accretion basically remained at the discretion of local authorities, it may be reasonable to assume that the general rate of commutation conformed to local productivity levels. Surcharges such as military taxes which, as emergency levies, constituted clear exceptions to this expected rule, but taxes of this nature generally did not become important before the Republican era.

Since the Imperial decree of permanent settlement in the early eighteenth century, a few common types of surtax had however been integrated into the main tax rates.[5] Local official records on

---

[1] Ping-ti Ho, Studies on the Population of China (Cambridge, Mass.: Harvard University Press, 1959), pp. 67-73.

[2] Ibid., p. 101.

[3] See S. R. Wagel, p. 365; and Morse, pp. 81-90.

[4] See Rural Reconstruction Commission, T'ien-fu fu-chia-shui t'iao-ch'a [A survey of land surtaxes] (Shanghai: Commercial Press, 1935), pp. 18-27.

[5] The two major types were hao-hsien and p'ing-yü (see ibid.; pp. 19-20).

actual tax collections thus contained these levies as part of the main tax, but a variety of other surcharges were usually not reported. In an important study, however, Wang Yeh-chien was able to estimate the actual rates of collection from a variety of historical sources and from a collection of provincial financial reports submitted to the central government prior to the 1911 revolution.[1] Though Wang's estimates of the actual tax rates were based on reports around 1908, the fact that land tax, unlike land value, was slow in response to improvement in land productivity in late Imperial China renders these data applicable to a much earlier period, or at least before 1900 when railway construction began to pick up its pace. But if land tax had indeed changed between 1900 and 1908 in response to land productivity or local economic development, such changes would likely have affected railway localities more than non-rail localities. The result could only increase the probability of rejecting the null hypothesis of no ex ante difference in favour of railway localities and thus add to the level of confidence if the null hypothesis is accepted.

Land tax data at the hsien level were compiled from the three provincial gazetteers used earlier in the population density test.[2] In addition, the actual tax rates on land which comprised various surtaxes were obtained from Wang for three other provinces: Honan, Shanhsi, and Anhuei,[3] which had relatively large railway mileages in the 1930s. These data were then subjected to a difference-of-means test using the t-statistic for the two sets of samples representing rail and non-rail localities. The results are shown in table 23. It can be seen that significant differences in land tax rates, whether expressed in official quotas or actual rates of collection,

---

[1] See Wang Yeh-chien, Land Taxation in Imperial China, 1750-1911 (Cambridge, Mass.: Harvard University Press, 1973); and idem, An Estimate of the Land-Tax Collection in China, 1753 and 1908 (Cambridge, Mass.: Harvard University Press, 1973).

[2] Namely, Shantung t'ung chih (1910), vols. 79-80; Chi-fu t'ung-chih (1884), vol. 94; and Chechiang t'ung-chih (1899), vols. 61-70. I have subtracted the ting quotas from the ti-ting totals (note: ti and ting tax referred to the land and poll tax [officially the commutation of labor services], respectively. The ting quota, however, was fixed early in the Ch'ing dynasty and, by the mid-eighteenth century, its incidence had been completely shifted to the land or land tax under a series of fiscal reforms called the single-whip system [see Ho, pp. 24-35; see also T'ien-fu fu-chia-shui t'iao-ch'a, p. 19]). But mostly these adjustments did not yield significant changes in the estimated rates. As for silver-cash exchange ratios and grain prices (rice and wheat in North and South China resp.) needed for conversion of money rates into tax in-kind (plus tribute grain levies), I have relied on Maritime Customs Reports and local histories.

[3] Wang, An Estimate of the Land-Tax Collection in China, tables 4, 5, and 15.

TABLE 23

TESTING LAND-TAX RATE DIFFERENCE IN RAIL AND NON-RAIL HSIEN

Province	Railway Hsien			Non-Railway Hsien			t_1^*	t_2^{**}	$t_s^{\#}$
	Total Number	Mean Rate I*	Mean Rate II**	Total Number	Mean Rate I*	Mean Rate II**			
Shantung	20	1.311	...	85	1.014	...	1.332	...	1.99
Hopei	33	0.904	...	79	0.995	...	0.703	...	1.98
Chechiang	22	2.135	...	53	2.897	...	1.590	...	1.99
Honan	36	1.012	2.163	64	1.155	2.381	0.841	0.710	1.99
Shanhsi	9	1.098	1.534	94	1.866	2.694	0.850	1.100	1.99
Anhuei	9	1.846	1.725	51	2.816	2.839	1.311	1.209	2.00

SOURCES: See text.

*This refers to land and grain tribute tax rates based on official records of collection (in kg. grain equivalents per mou).

**This refers to land and grain tribute tax rates based on the actual amount of collection (including surtaxes) (in kg. grain equivalents per mou).

#Critical value of the t-statistic at 0.05 level of confidence.

were detected in none of the provinces. As a matter of fact, the sign of difference between mean rates suggests that land tax rates were actually higher in non-rail communities in most of the provinces examined (Shantung was the only exception).

The t-test was also applied to the samples of rail and non-rail localities that appeared in figure 9 (chapter IV). Since these samples contain localities in different provinces, the estimated land tax rates should pertain to the same time period. For this purpose the collection rates pertaining to the period 1870-90 were calculated from local histories or privincial gazetteers.[1] As for the actual rate of collection, which included surtaxes seldom reported in the local histories, Wang's figures were used where they are available. But in a number of provinces, such as Chiangsu, most of Hopei, Shantung, and Hupei, such data were not available at the hsien level, so that average provincial surtax rates estimated by Wang have been applied to officially reported hsien quotas collected from local histories. Apparently this approach is not entirely appropriate, but the actual rates of collection served only as a secondary indicator in the present test; the primary indicator is the official rates. A few localities, however, had to be abandoned either because no relevant local history existed or because such histories, if they exist, were not available to this writer. Four

---

[1]Sources are as follows: Hopei: Chi-fu t'ung chih (1884), vol. 94; Shanhsi: Shanhsi t'ung chih (1890), vols. 58-61; Anhuei: Anhuei t'ung chih (1877), vol. 70; Hunan: Hunan t'ung chih (1885), vols. 50-52; Fuchien: Fuchien t'ung chih (1871), vols. 49-50; Chianghsi: Chianghsi t'ung chih (1881), vol. 84; Chechiang: Chechiang t'ung chih (1899), vols. 61-70; Yunnan: Yunnan t'ung chih (1894), vol. 6; Shantung: Shou-kuang hsien chih (1936), vol. 5; Ning-yang hsien chih (1879), vol. 7; Lai-yang hsien chih (1935), vol. 2; Wei-min hsien chih (1886), vol. 2; Fu-shan hsien chih kao (1931), vol. 13; En hsien chih (1909), vol. 4; Yi hsien chih (1904), vol. 13; Wei hsien hsiang-t'u chih (1907), vol. 2; Chi-ning chih-li chou hsü-chih (1926), vol. 4; Chi-mo hsien chih (1873), vol. 5; T'ai-an hsien chih (1929), vol. 2; An-ch'iu hsien chih (1920), vol. 5; Honan: Nan-yang hsien chih (1904), vol. 3; Yen-ch'eng hsien chih (1934), vol. 9; Shang-ch'iu hsien chih (1885), vol. 2; Lin-chang hsien chih (1874), vol. 3; Fu-hsiang hsien chih (1898), vol. 2; Chi hsien chin chih (1935), vol. 4; Lo-yang (1939), vol. 5; Ling-pao hsien chih (1876), vol. 3; angsu: Ch'ang-chao ho-chih-kao (1904), vol. 10; Ch'ing-ho hsien chih (1876), vol. 7; Yang-chou fu chih (1874), vol. 20; K'un-hsin liang-hsien hsü chih (1922), vol. 6; T'ai-chou fu chih (1827), vol. 9; Wu-hsi chin-kuei hsien chih (1881), vol. 9; Wu-chin yang-hu hsien chih (1879), vol. 2; Yen-ch'eng hsien chih (1895), vol. 4; Fu-ning hsien chih (1886), vol. 5; Hupei: Chung-chang hsien chih (1937), vol. 7; Ch'i-shui hsien chih (1880), vol. 4; Ying-ch'eng hsien chih (1882), vol. 3; Yün-meng hsien chih (1883), vol. 3; S ch an: Ch'ungch'ing chou chih (1877), vol. 2; F'ou-chou chih (1870), vol. 4; Mien-yang hsien chih (1933), vol. 3; Nei-chiang hsien-chih (1905), vol. 2; Sui-ning hsien chih (1879), vol. 6; Ta hsien chih (1933), vol. 11; K ang : Yung hsien chih (1897), vol. 9; Suiyuan: K'uei-sui hsien chih (1935), ching-cheng chih; Liaoning: Liao-chung hsien chih (1930), vol. 18; Shenhsi: Wei-nan hsien chih (1892), vol. 5; Kuangtung: Hsiang-shan hsien-chih (1879), vol. 7; Chao-ch'ing fu chih (1876), vol. 9; Chieh-yang hsien chih (1890), vol. 3; Ch'ü-chiang hsien chih (1875), vol. 12.

such localities, two each of the railway and non-railway samples, were in the Wheat Region, and two, one each in the two samples, were in the Rice Region. Test results of the remaining localities are given in table 24, which could not reject the null hypothesis of no ex ante productivity difference in the dichotomized samples at the 0.01 level of confidence in both crop regions.

Though some reservation should be made with regard to any conclusion drawn from these results because of the difficulty of ascertaining the exact extent to which land tax rates were indicative of actual land productivity differences, the present results apparently furnish further evidence in support of the hypothesis. In chapter V of the text, yield statistics of a number of commercial crops provide an alternative set of evidence which in fact indicates the possibility of unfavorable productivity conditions in many railway areas. Obviously this possibility is also implied by the above findings.

TABLE 24

TESTING LAND-TAX RATE DIFFERENCE IN RAIL AND NON-RAIL SAMPLE LOCALITIES

	Railway Sample			Non-Railway Sample				
Region	Sample Size	Mean Rate I*	Mean Rate II**	Sample Size	Mean Rate I*	Mean Rate II**	t_1^*	t_2^{**}
Wheat	25	0.995	1.769	21	1.100	1.813	0.83[a]	0.20[a]
Rice	14	2.050	2.118	47	3.243	3.160	0.22[a]	0.38[a]

SOURCES: See text

*This refers to land and grain tribute tax rates based on official records of collection (in kg. grain equivalents per mou).

**This refers to land and grain tribute tax rates based on the actual amount of collection, including surtaxes of various kinds (in kg. grain equivalents per mou).

[a] Null hypothesis of no difference between mean rates accepted at 0.01 level of confidence.

APPENDIX E

ESTIMATION OF THE AVERAGE DECLINE IN RURAL-URBAN TRANSFER
COSTS AND ITS CONTRIBUTION TO INCREASES IN FARM PRICES

In this appendix an estimation of the average reduction in the magnitude of rural-urban transfer costs will be attempted using both rural farm price time-series and wholesale agricultural price time-series at major terminal urban and export markets. Though reduction in rural transport costs may have been a result of local and regional transport developments without involving the construction of railways, the advent of rail transport was clearly instrumental in the contraction of economic distances in interregional commodity flows.

Consider the bundle of marketed agricultural commodities j ($j = 1, \ldots, m$) at any rural locality i ($i = 1, \ldots, n$). At equilibrium, farm prices tended to differ from those at the terminal urban and export market(s) by the total costs of transfer, which for simplicity is assumed to be constant for all farm produce between any rural locality and the terminal market. Thus

$$y_{ij} + x_i = z_{(i)j} \quad (i = 1,\ldots, n; \; j = 1,\ldots, m) \tag{1}$$

where y_{ij} = farm price of commodity j at locality i; x_i = transfer cost from i to the urban market; and $z_{(i)j}$ = the wholesale price of commodity j at the urban market (the subscript (i) is for identification purpose, referring to the fact that in this case we are talking about the relation between locality i and the urban market. Note that z_j is exogenously determined and does not vary with i in any specific time period). Define a set of quantity weights q_{ij} according to the relative importance of the commodity j at locality i.[1] Multiply (1) throughout by q_{ij} and sum the products over all commodities, we have

$$\sum_{j=1}^{m} q_{ij} y_{ij} + \sum_{j=1}^{m} q_{ij} x_i = \sum_{j=1}^{m} q_{ij} z_{(i)j} \tag{2}$$

Let the superscript t ($t = -k, \ldots, 0, \ldots, k$) be an index of time with unit in years and let $t = 0$ be the base or reference time

---

[1] The q_{ij}'s used by Buck were the percentages of the crop area occupied multiplied by the respective average yields per mou (see Land Utilization, p. 312). These same weights have been used in the following calculations.

period. By choosing the quantity weights q_{ij}^0 in (2) as reference weights for the price series (y^t; x^t; z^t) and dividing (2) throughout by the quantity $\sum_{j=1}^{m} q_{ij}^0 y_{ij}^0$ we obtain

$$\frac{\sum_{j=1}^{m} q_{ij}^0 y_{ij}^t}{\sum_{j=1}^{m} q_{ij}^0 y_{ij}^0} + \frac{\sum_{j=1}^{m} q_{ij}^0 x_i^t}{\sum_{j=1}^{m} q_{ij}^0 y_{ij}^0} = \frac{\sum_{j=1}^{m} q_{ij}^0 z_{(i)j}^t}{\sum_{j=1}^{m} q_{ij}^0 y_{ij}^0} \qquad (3)$$

It is easily seen that the first term on the left hand side of the equation, $\sum_{j=1}^{m} q_{ij}^0 y_{ij}^t / \sum_{j=1}^{m} q_{ij}^0 y_{ij}^0$, is simply the Laspeyre price index at locality i with t as the comparison time period. The relation specified by (3) can be used to measure the absolute change in the price indexes from the base period values at locality i:

$$(1 - L^t) + (x^0 - x^t) w = (1/g) (L_m^0 - L_m^t) \qquad (4)$$

for any i where L^t = Laspeyre farm price index at time t (note that $L^0 = 1$); $w = \sum_{j=1}^{m} q_{ij}^0 / \sum_{j=1}^{m} q_{ij}^0 y_{ij}^0$; $g = y^0 / z^0$ (here assumed to be constant for all j commodities in any particular locality i); and L_m^0 and L_m^t = Laspeyre agricultural price indexes at the urban market using quantity weights of locality i for the base and comparison time period, respectively. The absolute decline in total transfer cost for locality i is therefore $(x^0 - x^t)$ between time periods t=0 and t=t, and the percentage contribution of transfer cost reduction to the rise in farm price indexes is

$$d = \frac{(x^0 - x^t) w}{1 - L^t} \qquad (5)$$

In the present exercise, the base period is chosen to be 1930 and the start, as well as the comparison period, is the first year for which time-series data on local farm prices is available, with 1900 as the earliest comparison period if data for that year existed. $(1 - L^t)$ is computed from the time-series data in Buck for some 30 localities.[1] The value of the parameter g is judged to be around 0.8-0.9 for most of the localities covered by Buck's survey.[2] To estimate L_m, time-series data on the wholesale prices of ten major

---

[1] Data appear in Buck, LUSV, p. 149. I have also included Wusiang (Shanhsi) and Yenshan (Hopei) in the analysis. The price data for these two localities appear in Buck, "Price Changes in China," pp. 239-40.

[2] See Buck, Silver and Prices in China [Shanghai: Commercial Press, 1935], pp. 28-29, 34-35, 40-41.

marketed farm products, which were the main items included in Buck's calculation of the farm price indexes,[1] were collected from Customs trade reports and other sources,[2] with the assumption that Tientsin and Shanghai represented the main terminal urban and export markets in North and South China, respectively. The q_{ij}'s have been taken directly from Buck in computing the L_m's. Substituting the estimated index values into equation (4), the d values as defined by equation (5) were derived. From these results it was found that the average d, i.e. the percentage increase in local farm prices accounted for by the decline in transfer costs, was about 20.4% for North China, but only about 4% for South China. Apparently these findings conform to the differential rate of railway development in the two regions and also to the traditional reliance on a relatively low-cost system of internal water transport in a considerable part of South China.

---

[1] They are rice, tea, cotton, tobacco, seasame, wheat, millet, kaoliang, soybean, and peanut. Price series for corn, barley, and rapeseed are not available for the computation.

[2] I have used average export prices as surrogates for terminal urban market prices for a number of commodities. Data in this regard were obtained from Customs Trade Reports and Tang, Economic Study of Chinese Agriculture, pp. 374-80. These data have been supplemented by price data collected at the wholesale markets of Tientsin and Shanghai. The sources are Nankai Institute of Economics, Nankai chih-shu, pp. 70-73; Shanghai Economic Research Office of the Chinese Institute of Sciences, Shanghai wu-chia tzu-liao, p. 213; and China National Tariff Commission, Monthly Report on Prices and Price Indexes in Shanghai (Shanghai: Ministry of Finance), starting from 1919.

BIBLIOGRAPHY

Chinese Sources

Books

Bank of China, Economic Research Unit. Mi [Rice]. Nanking: Bank of China, 1937.

Buck, John Lossing. Hopei yen-shan hsien i-pai-wu-shih nung-chia chih ching-chi chi she-hui t'iao-ch'a [An economic and social survey of one hundred and fifty farm families in Yenshan Hsien, Hopei Province]. Nanking: Nanking University, Department of Agricultural Economics. No date of publication.

Bureau of Roads, Ministry of Communications. Chung-kuo ti kung-lu [Highways in China]. Nanking: National Economic Council, 1936.

Chang, Ch'i-yün. Chung-hua min-kuo ti-t'u chih [Atlas of the Republic of China]. 5 vols. Taipei: National War College, 1963.

Chang, Chih-i and Ou, Pao-san. Fuchien-sheng shih-liang chih yun-hsiao. [Grain marketing and transportation in Fuchien Province]. Shanghai: Commercial Press, 1938.

Chang, Hsiao-mei, comp. Kueichou ching-chi [The economy of Kueichou Province]. Shanghai: China National Economic Research Office, 1939.

Chang, Hsin-ch'eng. Chung-kuo hsien-tai chiao-t'ung shih [History of modern Chinese communications]. Shanghai: Commercial Press, 1937.

Chang, Hsin-i. Hopei-sheng nung-yeh kai-k'uang ku-chi [An estimate of the agricultural condition in Hopei Province]. Nanking: Statistical Bureau. No date of publication.

Chang, P'ei-kang. Chianghsi shih-liang wen-t'i [The grain problem in Chianghsi Province]. Shanghai: Institute of Social and Economic Research, 1935.

_____. Kuanghsi liang-shih wen-t'i [The grain problem in Kuanghsi Province]. Shanghai: Commercial Press, 1938.

_____. and Chang, Chih-i. Chechiang-sheng shih-liang chih yun-hsiao [Transportation and marketing of grains in Chechiang Province]. Ch'angsha: Commercial Press, 1939.

Ch'en, Hui. Chung-kuo t'ieh-lu wen-t'i [China's railway problems]. Peking: San-lien, 1955.

Ch'en, Pai-chuang. Ping-Han t'ieh-lu yen-hsien nung-ts'un ching-chi t'iao-ch'a [An economic survey of rural villages along the Peking-Hankow Railway]. Shanghai: Chiao-t'ung University Press, 1936.

Ch'en, P'ei-ta. Chin-tai chung-kuo ti-tso kai-shuo [Introduction to land rents in modern China]. 2nd ed. Peking: People's Press, 1963.

Ch'i, Li. Shan-Kan-Ning pien-ch'u shih-lu [True account of the Shenhsi-Kansu-Ninghsia border region]. Yenan: Chieh-fang-she. Reproduced by Center for Chinese Research Materials, Association of Research Libraries, Washington, D.C., 1969.

Chiao-t'ung University. Yuan-chung tu-mi ch'an-hsiao chih t'iao-ch'a [A survey of rice marketing in central Anhuei Province]. Shanghai: Chiao-t'ung University, 1936.

Chou, Chih-hua. Chung-kuo chung-yao shang-pin [Major commercial articles in China]. Shanghai: Hua-t'ung Book Co., 1930.

Chu, Chien-pang. Yangtze-chiang hang-yeh [River transport on the Yangtze River]. Shanghai: Commercial Press, 1937.

Ch'ü Chih-shang. Nung-ch'an-pin yun-hsiao yen-chiu ti fang-fa [Method of studying the transportation of agricultural commodities]. Peking: Social Research Institute, 1933.

Chu, Mei-yu. Chung-kuo cha-yeh [Tea in China]. Shanghai: Chung-hua Book Co., 1937.

Chun, Y. S., and Huang, Y. L. Chung-kuo hai-kuan t'ieh-lu chu-yao shang-pin liu-t'ung kai-k'uang [An overview of the flow of major commercial articles through China's customs passes and railways]. Nanking: Ministry of Industry, 1937.

CKNY. See Li Wen-Chih and Chang You-i.

Fong, Hsien-t'ing (H. D. Fong). Chung-kuo chih mien-fang-chih yeh [Cotton textile industry in China]. Shanghai: Commercial Press, 1934.

Foreign Trade Bureau, Ministry of Industry. Hua-sheng [Peanuts (in China)]. Shanghai: Commercial Press, 1940.

_____. Chih-ma [Sesame (in China)]. Shanghai: Commercial Press, 1940.

_____. Yen-yeh [Tobacco (in China)]. Shanghai: Commercial Press, 1940.

Hangchou-shih ching-chi t'iao-ch'a [An economic survey of the city of Hangchou]. 2 vols. Shanghai: Commercial Press, 1932.

Ho, Yang-ning. Cha-Sui-Meng-min ching-chi ti chieh-p'ou [An analytic dissection of the economy of the Chahar-Suiyuan-Mongol region]. Shanghai: Commercial Press, 1935.

Hsieh, Pin. Chung-kuo t'ieh-lu shih [A history of Chinese railways]. Shanghai: Chung-hua, 1929.

Hu, Ch'iu-chen. Nung-yeh ching-chi kai-lun [An introduction to agricultural economics]. Shanghai: Chung-hua, 1937.

Institute of Social and Economic Research. Tsung-chiang mi-shih t'iao-ch'a [A survey of the Tsung-chiang rice market]. Shanghai: Institute of Social and Economic Research, 1936.

Kao, Lu-ming. T'ieh-lu yun-chia chih yen-chiu [A study on railway freight rates]. Published by the author. No date of publication.

Li, Kuo-ch'i. Chung-kuo tsou-ch'i ti t'ieh-lu chiung-yung [History of the early Chinese railway development]. Taipei: Chung-hua, 1961.

Li, Wen-chih, and Chang, You-i, eds. Chung-kuo chin-tai nung-yeh shih tz'u-liao [Historical materials on agriculture in modern China]. 3 vols. Peking: San-lien, 1957.

Liu, Chun-hang. Yüeh-han t'ieh-lu chu-shao-tun yun-shen chih ti-chia wen-t'i [Problems of land prices along the Chuchou-Shaokuan section of the Yüeh-Han Railway]. Original by Chung-kuo ti-cheng yen-chiu-shou, 1936. Reproduced by Taipei: Cheng-wen-chu-pen-she, 1977.

Lü, P'ing-teng, ed. Ssuch'uan nung-ts'un ching-chi [The rural economy of Ssuch'uan Province]. Shanghai: Commercial Press, 1936.

Ministry of Agriculture and Commerce. Nung-shang t'ung-chi piao [Tables of agricultural and commercial statistics] no. 3, 1914; no. 4, 1915; no. 6, 1917; no. 7, 1918. Peking: Ministry of Agriculture and Commerce, 1916-22.

Ministry of Communications. Chiao-t'ung shih, Lu-cheng p'ien [Section on railways, history of communications (in China)]. 18 vols. Nanking: Ministry of Communications, 1930.

Ministry of Industry. Chung-kuo ching-chi nien-chien [Chinese economic yearbook]. Nanking: Government Printing Office, 1934.

Ministry of Railways. T'ieh-lu nien-chien [Railway yearbook]. 3 vols. Nanking: Ministry of Railways, 1933-36.

Nan, Pin-fong. Honan ch'an-yü-ch'ü chih t'iao-ch'a pao-kao [A survey report of the tobacco producing areas in Honan Province]. Nanking: Nanking University Press, 1934.

Nankai Institute of Economics. Nankai chih-shu tzu-liao hui-p'ien [A compilation of materials on the Nankai price indexes]. Peking: Statistical Press, 1958.

Nanking University, Department of Agricultural Economics. Yu, Ngo, Yuan, Kan shih-sheng chih mien-ch'an yun-hsiao [Cotton marketing in the four Provinces of Honan, Hupei, Anhuei and Chiangsu]. Nanking: Nanking University Press, 1936.

P'ang, Tze-yi, ed. Chung-kuo chin-tai shou-kung-yeh shih tzu-liao [Source materials on the modern history of Chinese handicrafts]. 4 vols. Peking: San-lien, 1957.

Pi, Yu-cheng, ed. Chung-kuo chin-tai t'ieh-lu-shih tzu-liao [Documents on the modern history of Chinese railways]. 3 vols. Peking: Chung-hua Book Co., 1963.

Rural Reconstruction Commission. T'ien-fu fu-chia-shui t'iao-ch'a [An investigation of land surtaxes]. Shanghai: Commercial Press, 1935.

Rural Reconstruction Commission. <u>Chechiang-sheng nung-ts'un t'iao-ch'a</u> [An investigation of rural Chechiang Province]. Shanghai: Commercial Press, 1934.

_____. <u>Honan-sheng nung-ts'un t'iao-ch'a</u> [An investigation of rural Honan Province]. Shanghai: Commercial Press, 1934.

_____. <u>Chiangsu-sheng nung-ts'un t'iao-ch'a</u> [An investigation of rural Chiangsu Province]. Shanghai: Commercial Press, 1934.

_____. <u>Shenhsi-sheng nung-ts'un t'iao-ch'a</u> [An investigation of rural Shenhsi Province]. Shanghai: Commercial Press, 1934.

_____. <u>Kuanghsi-sheng nung-ts'un t'iao-ch'a</u> [An investigation of rural Kuanghsi Province]. Shanghai: Commercial Press, 1935.

_____. <u>Yünnan-sheng nung-ts'un t'iao-ch'a</u> [An investigation of rural Yünnan Province]. Shanghai: Commercial Press, 1935.

Shanghai Economic Research Office of the Chinese Institute of Sciences. <u>Shanghai chieh-fang ch'ien-hou wu-chia tzu-liao hui-p'ien</u> [A compilation of price materials for pre- and post-liberation Shanghai]. Shanghai: People's Press, 1958.

Shanghai te-pieh-shih she-hui-chu. <u>Shanghai chih kung-yeh</u> [The industries of Shanghai]. Shanghai: Chung-hua Book Co., 1930.

Shen Pao. <u>Shen Pao nien-chien</u> [Almanac of Shen Pao]. Shanghai: Shen Pao Press, 1933.

Shen, Tsou-t'ing. <u>T'ieh-lu wen-t'i tu-lun-chi</u> [Essays on the problems of railways]. Shanghai: Commercial Press. No date of publication.

Shen, Tsung-han. <u>Chung-kuo nung-yeh t'zu-yuan</u> [Agricultural resources of China]. 3 vols. Taipei: Chung-hua-wen-hua chu-pan-shih-yeh wei-yuan-hui, 1953.

Ting, Wen-chiang, ed. <u>Chung-hua min-kuo hsin-ti-t'u</u> [A new atlas of the Republic of China]. 2nd ed. Shanghai: Shen Pao Press, 1934.

Wang, Chien-hsin, tr. <u>Chung-kuo nung-shu</u> [Agricultural handbook of China]. 2 vols. Shanghai: Commercial Press, 1934.

Wang, Ch'in-yü. <u>Chin-tai Chung-kuo ti tao-lu chien-sheh</u> [Road and railway construction in modern China]. Hong Kong: Lung-man, 1969.

Wang, Ching-wang, tr. <u>Chung-kuo hsi-pei chih ching-chi chuang-k'uang</u> [The economic conditions of China's Northwest]. Shanghai: Commercial Press, 1933.

Wang, Kwang. <u>Chung-kuo shiu-yun chih</u> [River shipping in China]. Taipei: Chung-hua ta-tien p'ien-yin-hui, 1966.

Yang, Chun-chang. <u>Shang-kuei t'ieh-lu yu-shen ti-chia chih yen-chiu</u> [A study of land prices along the Hunan-Kueichou Railway]. Chung-kuo ti-cheng yen-chiu chung-han, 1938. Reproduced by Cheng-wen chu-pen-she, Taipei, 1977.

Yang, Hsiang-nien. <u>T'ieh-lu ching-chi yu t'sai-cheng</u> [Railway economics and finance]. Shanghai: Commercial Press, 1944.

Yang, Ming-sung. Chin che-nien wo-kuo she-shan-sheng wu-she-chiu-chü shang-ts'un wu-chia t'iao-ch'a [Survey of village commodity prices in 59 localities and 13 provinces in China in the past seven years]. Jungchiang, Scuch'uan: National agricultural Research Bureau, Ministry of Agriculture and Forestry, 1941.

Yen, Ch'ung-p'ing. Chung-kuo chin-tai ching-chi-shih t'ung-chi [Statistics on modern Chinese economic history]. Peking: Scientific Press, 1955.

_____. Chung-kuo mien-fang-chih shih-kao [Draft history of Chinese textiles]. Peking: K'o-hsüeh ch'u-pan-she, 1955.

Yu, Yen. Tsui-chin san-nien t'ieh-lu chien-ti yun-chia shu lueh [A brief account of the reduction of freight rates in the past three years]. Nanking: Ministry of Railways, 1935.

Articles

Chang, Po. "Tung-chi ching-chi-sha chih t'ieh-lu yun-chia cheng-ts'e" [Railway freight pricing policy in a regulated economy]. Chiao-t'ung tsa-chih 2 (January 1934):1-10.

Cheng, Ching-lieh. "Wo-kuo t'ieh-lu ying-yeh yung-k'uan chih kai-k'uang chi chi chen-li chih kuang-chien" [A study of the operating expenses of Chinese railways and an overview of its improvement]. Chiao-t'ung tsa-chih 1 (January 1933):72-82.

Chiang, Wu-chin. "Min-chuan chih yun-hsiao cheng-pen" [The cost of transportation by river junks]. Chiao-t'ung tsa-chih 3 (January 1935):13-26.

Fung, Ho-fa. "Chung-kuo nung-ch'an-wu ti yuan-shih shih-ch'ang" [Primary markets of agricultural products in China]. Chung-kuo-nung-ts'un 3 (December 1934):15-31.

Li, Wen-chih. "Land Rents, Commercial Capital, High Interest Rates, and the Peasants' Standard of Living in the Ch'ing Period prior to the Opium War." In Chung-kuo tz'u-pen-chu-i ming-ya wen-t'i t'ao-lun chi, pp. 609-56. Peking: San-lien, 1957.

Sun, Hsiao-ts'un. "K'uan-yu Chung-kuo mi-ku shang-pin-hua ti e-ke fen-shih" [An analysis of the commercialization of rice in China]. Chung-kuo-nung-ts'un 1 (December 1928):683-93.

Local Chinese Histories

Local Chinese Histories
(Gazetteers)(Provincial)

Anhuei t'ung-chih (1877 ed.)

Chechiang t'ung chih (1899 ed.)

Chi-fu t'ung chih (1884 ed.)

Fuchien t'ung-chih (1871 ed.)

Hunan t'ung chih (1885 ed.)

Chianghsi t'ung chih (1881 ed.)

Shanhsi t'ung chih (1890 ed.)

Shantung t'ung chih (1910 ed.)

Yunnan t'ung chih (1894 ed.)

Local Chinese Histories
(Gazetteers)(fu, chou and hsien)

An-ch'iu hsien chih (1920 ed.)(Shantung)

Ch'ang-chao ho-chih-kao (1904 ed.)(Chiangsu)

Chao-ch'ing fu chih (1876 ed.)(Kuangtung)

Chi hsien chin chih (1935 ed.)(Honan)

Chi-mo hsien chih (1873 ed.)(Shantung)

Chi-ning chih-li-chou hsü-chih (1926 ed.)(Shantung)

Ch'i-shui hsien chih (1880 ed.)(Hupei)

Chieh-yang hsien chih (1890 ed.)(Kuangtung)

Ch'ing-ho hsien chih (1876 ed.)(Chiangsu)

Ch'ü-chiang hsien chih (1875 ed.)(Kuangtung)

Chung-chang hsien chih (1937 ed.)(Hupei)

Ch'ung-ching chou chih (1877 ed.)(Ssuch'uan)

En-hsien chih (1909 ed.)(Shantung)

F'ou-chou chih (1870 ed.)(Ssuch'uan)

Fu-hsiang hsien chih (1898 ed.)(Honan)

Fu-ning hsien chih (1886 ed.)(Chiangsu)

Fu-shan hsien chih kao (1931 ed.)(Shantung)

Hsiang-shan hsien chih (1879 ed.)(Kuangtung)

K'uei-sui hsien chih (1935 ed.)(Suiyuan)

K'un-hsin liang-hsien hsü chih (1922 ed.)(Chiangsu)

Liao-chung hsien chih (1930 ed.)(Liaoning)

Lin-chang hsien chih (1874 ed.)(Honan)

Ling-pao hsien chih (1876 ed.)(Honan)

Lo-yang (1939 ed.)(Honan)

Mien-yang hsien chih (1933 ed.)(Ssuch'uan)

Nan-yang hsien chih (1904 ed.)(Honan)

Nei-chiang hsien chih (1905 ed.)(Ssuch'uan)

Ning-yang hsien chih (1879 ed.)(Shantung)

Shang-ch'iu hsien chih (1885 ed.)(Honan)

Shou-kuang hsien chih (1936 ed.)(Shantung)

Sui-ning hsien chih (1879 ed.)(Ssuch'uan)

Ta hsien chih (1933 ed.)(Ssuch'uan)

T'ai-an hsien chih (1929 ed.)(Shantung)

T'ai-chou fu chih (1827 ed.)(Chiangsu)

Wei hsien hsiang-t'u chih (1907 ed.)(Shantung)

Wei-min hsien chih (1886 ed.)(Shantung)

Wei-nan hsien chih (1892 ed.)(Shenhsi)

Wu-chin yang-hu hsien chih (1879 ed.)(Chiangsu)

Wu-hsi chin-kuei hsien chih (1881 ed.)(Chiangsu)

Yang-chou fu chih (1874 ed.)(Chiangsu)

Yen-ch'eng hsien chih (1895 ed.)(Chiangsu)

Yen-ch'eng hsien chih (1934 ed.)(Honan)

Yi hsien chih (1904 ed.)(Shantung)

Ying-ch'eng hsien chih (1882 ed.)(Hupei)

Yün-meng hsien chih (1883 ed.)(Hupei)

Yung hsien chih (1897 ed.)(Kuanghsi)

English Sources

Books

Allen, G. C. and Donnithorne, A. G. Western Enterprise in Far Eastern Economic Development. New York: Kelley, 1968.

Anderson, George E. Cotton Goods in China. Washington, D.C.: U.S. Bureau of Manufactures, 1911.

Arnold, Julean. Commercial Handbook of China. 2 vols. Washington D.C.: Government Printing Office, 1919-20.

Behrman, Jere R. Supply Response in Underdeveloped Agriculture. Amsterdam: North-Holland, 1968.

Buck, John Lossing. Chinese Farm Economy: A Study of 2866 Farms in 17 Localities and Seven Provinces. Chicago: University of Chicago Press, 1930.

_____. Silver and Prices in China. Shanghai: Commercial Press, 1935.

_____. Land Utilization in China. Nanking: Nanking University Press, 1937.

Buck, John Lossing. *Land Utilization in China: Statistical Volume*. Chicago: University of Chicago Press, 1937.

Chang, John K. *Industrial Development in Pre-Communist China*. Chicago: Aldine, 1969.

Chang, Yin-t'ang. *The Economic Development and Prospects of Inner Mongolia*. Shanghai: Commercial Press, 1933.

Chao, Kang. *The Development of Cotton Textile Production in China*. Cambridge, Mass.: Harvard University Press, 1977.

Chen, Han-seng. *Landlord and Peasant in China*. New York: International Publishers, 1936.

──────. *Industrial Capital and Chinese Peasants*. Shanghai: Kelley & Walsh, 1939.

Cheng, Ming-ju. *The Influence of Communications on the Economic Future of China*. London: G. Routledge & Sons, 1930.

Cheng, Yu-kwei. *Foreign Trade and Industrial Development of China*. Washington, D.C.: Washington University Press, 1956.

Chiang, K.N. *China's Struggle for Railway Development*. New York: John Day, 1943.

Chorley, Richard, and Haggett, Peter, eds. *Models in Geography*. London: Methuen, 1967.

Chow, Chuen-tyi. "China's Internal Transport Problem: The Case of the Railways' First Century, 1866-1966." Ph.D. dissertation, Michigan State University, 1972.

Chu, T. H. *Tea Trade in Central China*. Shanghai: Kelley & Walsh, 1936.

Chuan, Han-seng, and Kraus, Richard. *Mid-Ch'ing Rice Markets and Trade*. Cambridge, Mass.: Harvard University Press, 1975.

Clark, Colin, and Haswell, Margaret. *The Economics of Subsistence Agriculture*. New York: Macmillan, 1967.

Clark, Grover. *Economic Rivalries in China*. New Haven: Yale University Press, 1932.

Condliffe, John Bell. *China To-day: Economic*. Boston: World Peace Foundation, 1932.

Cowan, C. D., ed. *The Economic Development of China and Japan*. New York: Praeger, 1964.

Cressey, George B. *Land of the 500 Millions*. New York: McGraw Hill, 1955.

de Fellner, Frederick V. *Communications in the Far East*. London: P. S. King & Sons, 1934.

Edkins, Joseph. *Banking and Prices in China*. Shanghai: Presbyterian Mission Press, 1905.

Elvin, Mark. *The Pattern of the Chinese Past*. Stanford: Stanford University Press, 1973.

Fairbank, John K.; Reischauer, Edwin O.; and Craig, Albert M. East Asia: The Modern Transformation. Boston: Houghton, 1965.

Fei, Hsiao-t'ung. Peasant Life in China. London: Routledge & Kegan Paul, 1939.

Fei, Hsiao-t'ung, and Chang, Chih-i. Earthbound China. London: Routledge & Kegan Paul, 1945.

Feuerwerker, Albert. The Chinese Economy, ca. 1870-1911. Ann Arbor: University of Michigan, Center for Chinese Studies, 1969.

_____. The Chinese Economy, 1912-1949. Ann Arbor: University of Michigan, Center for Chinese Studies, 1968.

Fishlow, Albert. American Railroads and the Transformation of the Ante-Bellum Economy. Cambridge, Mass.: Harvard University Press, 1965.

Fogel, Robert W. Railroads and American Economic Growth. Baltimore: Johns Hopkins University Press, 1964.

Fong, Hsien-t'ing (H. D. Fong). Rural Industries in China. Tientsin: Chihli Press, 1933.

Foreign Trade Bureau, Ministry of Industry. Statistics of China's Foreign Trade by Ports, 1900-1933. Part I: Central Ports. Shanghai: Commercial Press, 1935.

Fromm, G., ed. Transport Investment and Economic Development. Washington, D.C.: Brookings Institution, 1965.

Gamble, Sidney D. Ting Hsien: A North China Rural Community. Stanford: Stanford University Press, 1954.

_____. North China Villages: Social, Political and Economic Activities before 1933. Berkeley: University of California Press, 1963.

Garrison, William L., and Marble, Duane F. The Structure of Transportation Networks. Washington, D.C.: U.S. Department of Commerce, Office of Technical Services, 1961.

Gould, Peter R. The Development of the Transportation Pattern in Ghana. Department of Geography Studies in Geography No. 5. Evanston, Ill.: Northwestern University, 1960.

Greenberg, M. British Trade and the Opening of China, 1800-1842. Cambridge: Cambridge University Press, 1951.

Hayami, Y. and Ruttan, V. W. Agricultural Development: An International Perspectives. Baltimore: Johns Hopkins University Press, 1971.

Haswell, Margaret. Tropical Farming Economics. London: Longmans, 1973.

Heady, Earl O., and Dillon, J. L. Agricultural Production Functions. Ames: Iowa State University Press, 1961.

Herrmann, Albert. Historical and Commercial Atlas of China. Chicago: Aldine, 1966.

Hinton, H. C. The Grain Tribute System of China, 1845-1911. Cambridge, Mass.: Harvard University Press, 1956.

Ho, Ping-ti. *Studies on the Population of China, 1368-1953.* Cambridge, Mass.: Harvard University Press, 1959.

Hou, Chi-ming. *Foreign Investment and Economic Development in China, 1840-1937.* Cambridge, Mass.: Harvard University Press, 1965.

Hoyle, B. S. *Transport and Development.* New York: Barnes and Noble, 1973.

Hsiao, Liang-lin. *China's Foreign Trade Statistics, 1864-1949.* Cambridge, Mass.: Harvard University Press, 1974.

Hsu, Mongton C. *Railway Problems in China.* New York: Columbia University Press, 1915.

Jing, Su, and Lou, Lun. *Landlord and Labor in Late Imperial China: Case Studies from Shantung.* Translated by E. Wilkinson. Cambridge, Mass.: Harvard University Press, 1978.

Johnston, J. *Econometric Methods.* 2nd ed. New York: McGraw Hill, 1972.

Jones, E. L., and Woolf, S. J., eds. *Agrarian Change and Economic Development: The Historical Problems.* London: Methuen, 1969.

Kent, Percy. *Railway Enterprise in China.* London: Edward Arnold, 1907.

King, Frank. *A Concise Economic History of China.* New York: Praeger, 1967.

Kulp, D. H. *Country Life in South China.* New York: Columbia University Press, 1925.

Kuznets, Simon. *Six Lectures on Economic Growth.* New York: Free Press, 1959.

Lee, En-han. *China's Struggle for Railway Autonomy.* Singapore: Singapore University Press, 1977.

Lele, Uma. *The Design of Rural Development.* Baltimore: Johns Hopkins University Press, 1975.

Leung, C. K. *China: Railway Patterns and National Goals.* Chicago: University of Chicago Department of Geography Research Paper No. 196, 1980.

Lieu, D. K. *China's Industries and Finance.* Shanghai: Chinese Government Bureau of Economic Information, 1927.

Liu, Ta-chung, and Yeh, Kung-chia. *The Economy of the Chinese Mainland.* Princeton: Princeton University Press, 1965.

Liu, Ts'ui-jung. "Trade on the Han River and Its Impact on Economic Development, ca. 1800-1911." Ph.D. dissertation, Harvard University, 1974.

McAlpin, M. B. "The Impacts of Railroads on Agriculture in India, 1860-1900: A Case Study of Cotton Cultivation." Ph.D. dissertation, University of Wisconsin at Madison, 1973.

Mallory, W. H. *China: Land of Famine.* New York: American Geographical Society, 1926.

Malone, C. B., and Taylor, J. B. The Study of Chinese Rural Economy. Peking: Peking United International Famine Relief Committee Reports Series B #10, 1924.

Maritime Customs, Republic of China Decennial Reports, 1902-1911. Shanghai: Maritime Customs, 1912.

──────. Decennial Reports, 1912-21. Shanghai: Maritime Customs, 1922.

──────. Decennial Reports, 1921-1931. Shanghai: Maritime Customs, 1932.

──────. Reports and Returns of Trade for Treaty Ports (various years). Shanghai: Maritime Customs, 1879-1930.

Mellor, John W. The Economics of Agricultural Development. Ithaca: Cornell University Press, 1966.

Mood, A. M., and Graybill, F. A. Introduction to the Theory of Statistics. 2nd ed. New York: McGraw Hill, 1963.

Morse, H. B. The Trade and Administration of China. London: Longmans, 1913.

Moyer, Reed, and Hollander, Stanley C., eds. Markets and Marketing in Developing Countries. Homewood, Ill.: Irwin, 1968.

Murphey, Rhoads. Shanghai: Key to Modern China. Cambridge, Mass.: Harvard University Press, 1953.

Myers, Ramon H. The Chinese Peasant Economy: Agricultural Development in Hopei and Shantung, 1890-1949. Cambridge, Mass.: Harvard University Press, 1970.

National Tariff Commission, Republic of China, Prices and Price Indexes in Shanghai. Shanghai: National Tariff Commission, 1937.

O'Brien, Patrick. The New Economic History of the Railways. New York: St. Martin's Press, 1977.

O'Connor, A. M. Railways and Development in Uganda: A Study in Economic Geography. Nairobe: Oxford University Press, 1965.

Owen, W. Distance and Development: Transport and Communications in India. Washington, D.C.: Brookings Institution, 1968.

Perkins, Dwight. Agricultural Development in China, 1368-1968. Chicago: Aldine, 1969.

Potter, Jack. Capitalism and the Chinese Peasants. Berkeley: University of California Press, 1968.

Price, E. B. Fukien: A Study of a Province in China. Shanghai: Presbyterian Mission Press, 1925.

Pusey, James. Wu-han: Attacking the Present from the Past. Cambridge, Mass.: Harvard University Press, 1969.

Rawski, Evelyn S. Agricultural Change and the Peasant Economy of South China. Cambridge, Mass.: Harvard University Press, 1972.

Reed, P. W. *The Economics of Public Enterprises.* London: Butterworths, 1973.

Remer, C. F. *Foreign Investments in China.* New York: Macmillan, 1933.

Reynolds, Bruce. "The Impact of Trade and Foreign Investment on Industrialization: Chinese Textiles, 1875-1931." Ph.D. dissertation, University of Michigan, 1975.

Roll, Charles Robert Jr. *The Distribution of Rural Income in China: A Comparison of the 1930s and the 1950s.* Cambridge, Mass.: Harvard University Press, 1975.

Rowe, David N. *China Among the Powers.* New York: Harcourt, Brace and Co., 1945

Rozman, Gilbert. *Urban Networks in Ch'ing China and Tokugawa Japan.* Princeton: Princeton University Press, 1973.

Schrecker, John E. *Imperialism and Chinese Nationalism: Germany in Shantung.* Cambridge, Mass.: Harvard University Press, 1971.

Schultz, T. W. *Transforming Traditional Agriculture.* New Haven: Yale University Press, 1964.

_____. *Economic Growth and Agriculture.* New York: McGraw Hill, 1968.

Smith, C. A. Middleton. *The British in China.* London: Constable, 1920.

Southworth, H. M., and Johnston, Bruce F., eds. *Agricultural Development and Economic Growth.* Ithaca: Cornell University Press, 1969.

Stringer, H. *The Chinese Railway System.* Tientsin: Tientsin Press Ltd., 1925.

Sun, Kungtu C. *The Economic Development of Manchuria in the First Half of the 20th Century.* Cambridge, Mass.: Harvard University Press, 1973.

Sun, Yat-sen. *The International Development of China.* New York: G. P. Putnam, 1922.

Taaffe, Edward J., and Gauthier, Howard L. Jr. *Geography of Transportation.* Englewood Cliffs, N.J.: Prentice-Hall, 1973.

Tang, Chi-yu. *An Economic Study of Chinese Agriculture.* No publisher, 1924.

Tawney, R. H., ed. *Agrarian China.* London: Allen & Unwin, 1939.

_____. *Land and Labor in China.* New York: Octagon, 1964.

Tayler, J. B. *Farm and Factory in China.* London: Student Christian Movement, 1928.

Torgasheff, Boris P. *China as a Tea Producer.* Shanghai: Commercial Press, 1926.

Tregear, T. R. *An Economic Geography of China.* New York: Elsevier, 1970.

Vinacke, Harold M. *Problems of Industrial Development in China*. Princeton: Princeton University Press, 1926.

Wagel, S. R. *Finance in China*. Shanghai: North China Daily News and Herald, 1914.

Wang, Yeh-chien. *An Estimate of the Land-Tax Collection in China, 1753 and 1908*. Cambridge, Mass.: Harvard University Press, 1973.

_____. *Land Taxation in Imperial China, 1750-1911*. Cambridge, Mass.: Harvard University Press, 1973.

Watson, A., tr. *Transport in Transition: The Evolution of Traditional Shipping in China*. Michigan Abstracts of Chinese and Japanese Works on Chinese History No. 3. Ann Arbor: University of Michigan, 1972.

Whitney, Joseph. *China: Area, Administration, and Nation Building*. Department of Geography Research Paper No. 123. Chicago: University of Chicago, 1970.

Yang, C., comp. *Statistics of China's Foreign Trade during the Last Sixty-Five Years*. Shanghai: National Research Institute of Social Science, Academia Sinica, 1931.

Yang, C. K. *A Chinese Village in Early Communist Transition*. Cambridge, Mass.: Technology Press, 1959.

Yang, Ching-k'un. *A North China Local Market Economy: A Summary of a Study of Periodic Markets in Chowping Hsien, Shantung*. New York: International Secretariat, Institute of Pacific Relations, 1944.

Yang, L. S. *Money and Credit in China*. Cambridge, Mass.: Harvard University Press, 1952.

Yang, M. C. *A Chinese Village - Taitou, Shantung Province*. New York: Columbia University Press, 1945.

Yotopolous, Pan A., and Nugent, Jeffrey B. *Economics of Development*. New York: Harper & Row, 1976.

Young, Arthur. *China's Nation-Building Effort*, 1927-37. Stanford: Hoover Institution, 1971.

Articles

Abbott, J. C. "The Role of Marketing in the Development of Backward Economies." *Journal of Farm Economics* 44 (May 1962):349-62.

Abramovitz, Moses. "The Economic Characteristics of Railroads and the Problem of Economic Development." *Far Eastern Quarterly* 14 (June 55):169-78.

Brown, Shannon R. "The Transfer of Technology to China in the Nineteenth Century: The Role of Direct Foreign Investment." *Journal of Economic History* 39 (March 1979):181-98.

Buck, John Lossing. "Price Changes in China." *Journal of the American Statistical Association* 25 (June 1925): 40-43.

Chang, L. L. "Farm Prices in Wuchin, Kiangsu." *Chinese Economic Journal* 10 (June 1932):449-512.

Chen, Chunjen C. "Tobacco-Growing in Shantung." Chinese Economic Journal 10 (January 1932):37-44.

Chiao, C. M. and Buck, J. L. "The Composition and Growth of Rural Groups in China." Chinese Economic Journal 2 (March 1928):219-35.

"Chinese Flour Industry." Chinese Economic Journal 8 (February 1931): 106-12.

Chou, Shun-hsin. "Railway Development and Economic Growth in Manchuria." China Quarterly 45 (January/March 1971):57-84.

Ch'üan, Han-seng. "Production and Distribution of Rice in Southern Sung." In Chinese Social History, pp. 222-33. Edited by E-tu Zen Sun. Washington, D.C.: American Council of Learned Societies, 1956.

Coatsworth, John. "Indispensable Railroads in a Backward Economy: The Case of Mexico." Journal of Economic History 39 (December 1979):939-60.

"Cotton Production in Chekiang." Chinese Economic Journal 13 (September 1933):258-70.

"Cotton Production in Hupei." Chinese Economic Journal 13 (September 1933):356-63.

Dernberger, Robert F. "The Role of the Foreigner in China's Economic Development, 1840-1949." In China's Modern History in Historical Perspective, pp. 19-48. Edited by Dwight H. Perkins. Stanford: Stanford University Press, 1975.

Dittmer, C. G. "An Estimate of the Chinese Standard of Living in China." Quarterly Journal of Economics 33 (January 1918):107-28.

Dittrich, Scott R., and Myers, Ramon H. "Resource Allocation in Traditional Agriculture: Republican China, 1937-1940." Journal of Political Economy 79 (December 1971):887-96.

Dixit, A. K. "Marketable Surplus and Dual Development." Journal of Economic Theory 1 (August 1969):203-19.

"Famines in Shantung." Chinese Economic Journal 2 (January 1928): 36-43.

Fang, Fu-an. "Ten Years of Road Building in China--A Statistical Study." Chinese Economic Journal 6 (May 1930):542-57.

_____. "Chinese Cotton Industry, 1930." Chinese Economic Journal 7 (November 1930):1197-1230.

"Flour Industry in Shantung." Chinese Economic Journal 15 (September 1934):328-37.

Fogel, Robert W. "Notes on the Social Savings Controversy." Journal of Economic History 39 (March 1979):1-54.

Gamble, Sidney D. "Daily Wages of Unskilled Chinese Laborers, 1807-1902." Far Eastern Quarterly 3 (November 1943):41-73.

Gauthier, Howard L. "Transportation and the Growth of the Sao Paulo Economy." Journal of Regional Science 8 (February 1968):77-84.

Ginsburg, Norton S. "Manchurian Railway Development." China Quarterly 8 (August 1949):398-411.

Ginsburg, Norton S. "From Colonialism to National Development: Geographical Perspectives on Patterns and Policies." Annals of the Association of American Geographers 63 (March 1973):1-22.

Griliches, Zvi. "Measuring Inputs in Agriculture: A Critical Survey." Journal of Farm Economics 42 (December 1960):1411-27.

_____. "Specification and Estimation of the Aggregate Agricultural Production Function from Cross-Section Data." Journal of Farm Economics 45 (May 1963):419-28.

Guenther, Stein. "China's Internal Transport System." Far Eastern Survey 20 (October 1943):209.

Habakkuk, H. J. "Historical Explanation of Economic Development." In Problems in Economic Development, chapter 6. Edited by E. A. G. Robinson. London: Macmillan, 1965.

Herdt, Robert W. "A Disaggregate Approach to Aggregate Supply." American Journal of Agricultural Economics 4 (November 1970): 512-20.

Hirschman, Albert O. "A Generalized Linkage Approach to Development." In Essays in Economic Development and Cultural Change in Honor of Bert F. Hoselitz, pp. 67-98. Edited by Manning Nash. Chicago: University of Chicago Press, 1977.

Ho, Franklin. "Price and Price Indexes in China." Chinese Economic Journal 5 (May 1927):429-63.

_____. "Prices and Price Fluctuations in North China, 1913-1929." Chinese Social and Political Science Review 5 (October 1929): 349-58.

Hoch, Irving. "Estimation of Production Function Parameters Combining Time-Series and Cross-Section Data." Econometrica 30 (January 1962):34-53.

Hopper, W. David. "Allocation Efficiency in a Traditional Indian Agriculture." Journal of Farm Economics 47 (August 1965):611-25.

Hornby, J. M. "Investment and Trade Policy in the Dual Economy." Economic Journal 78 (March 1968):96-107.

Hunter, Holland. "Transport in Soviet and Chinese Development." Economic Development and Cultural Change 14 (January 1965): 71-72.

_____. "Chinese and Soviet Transport for Agriculture." Pittsburg: University of Pittsburg, University Center for International Studies, 1974. Mimeographed.

Hurd, John II. "Railways and the Expansion of Markets in India, 1861-1921." Explorations in Economic History 12 (September 1975):263-88.

_____. "The Economic Impacts of Railways in India, 1853-1947." Paper presented at the Economic History Workshop, University of Chicago, June 1976.

Hymer, Stephen. "The Decline of Rural Industry under Export Expansion." *Journal of Economic History* 30 (March 1970):51-73.

Hymer, Stephen, and Resnick, Stephen. "A Model of an Agrarian Economy with Non-agricultural Activities." *American Economic Review* 59 (September 1969):493-506.

Johnston, Bruce F. "Agriculture and Structural Transformation in Developing Countries: A Survey of Research." *Journal of Economic Literature* 8 (June 1970):369-404.

Johnston, Bruce F., and Mellor, John W. "The Role of Agriculture in Economic Development." *American Economic Review* 51 (September 1961):566-93.

Jones, Donald W. "Rents in an Equilibrium Model of Land Use." *Annals of the Association of American Geographers* 68 (June 1978):205-13.

_____. "The Price Theory of Thunen's Theory of Rent: Assumptions and Implications of a Familiar Model." Department of Geography Discussion Paper. Chicago: University of Chicago, 1979.

Katzman, Martin T. "The von Thunen Paradigm, the Industrial-Urban Hypothesis, and the Spatial Structure of Agriculture." *American Journal of Agricultural Economics* 56 (November 1974):683-96.

Kansky, K. J. *Structure of the Transport Network*. Chicago: University of Chicago, Department of Geography Research Paper No. 84, 1963.

Kung, H. O. "The Growth of Population in Six Large Chinese Cities." *Chinese Economic Journal* 22 (March 1937):301-14.

Lattimore, Owen. "China and the Barbarians." In *Empire in the East*, pp. 23-32. Edited by Joseph Barnes. New York: Doubleday, 1934.

Lefeber, Louis. "Transport Development and Regional Growth." In *Transport Investment and Economic Development*, pp. 108-22. Edited by Gary Fromm. Washington, D.C.: Brookings Institution, 1965.

Lewis, Bradley G. "Economic Consequences of the Railroads Reconsidered." Economic History Workshop Discussion Paper No. 7677-28. Chicago: University of Chicago, June 10, 1977.

Liang, Ernest P. "Market Accessibility and Agricultural Development in Prewar China." *Economic Development and Cultural Change* (in press).

McCabe, James L. "Regional Product Price Differences in Less Developed Countries." *Journal of Political Economy* 85 (September 1977):549-68.

Mellor, John W. "The Functions of Agricultural Prices in Economic Development." *Indian Journal of Agricultural Economics* 23 (January-March 1968):23-28.

_____. "Accelerated Growth in Agriculture Production and the Sectoral Transfer of Resources." *Economic Development and Cultural Change* 22 (October 1973):1-16.

Metzer, Jacob. "Railroad Development and Market Integration: The Case of Tsarist Russia." Journal of Economic History 34 (September 1974):529-49.

Metzer, Jacob. "Railroads in Tsarist Russia: Direct Gains and Implications." Explorations in Economic History 13 (January 1976): 85-112.

Mitsugui, Matsuda. "Periodic Markets in North China during the Sung, Ming and Ch'ing Periods." In Markets in China during the Sung, Ming and Ch'ing Periods, pp. 109-42. Edited by Shiba Yashinori and Yamane Yukio. Honolulu: University of Hawaii, East-West Center, 1967.

Murphey, Rhoads. "The Treaty Ports and China's Modernization." In The Chinese City between Two Worlds, pp. 17-72. Edited by Mark Elvin and G. William Skinner. Stanford: Stanford University Press, 1974.

Myers, Ramon H. "Agrarian Policy and Agricultural Transformation: Mainland China and Taiwan, 1895-1945." Journal of the Institute of Chinese Studies, Chinese University of Hong Kong 3 (1970):521-42.

_____. "The Commercialization of Agriculture in Modern China." In Economic Organization in Chinese Society, pp. 173-91. Edited by W. E. Wilmott. Stanford: Stanford University Press, 1972.

_____. "Wheat in China--Past, Present and Future." China Quarterly No. 74 (June 1978):297-333.

Nicholls, William H. "Development in Agrarian Economies: The Role of Agricultural Surplus, Population Pressures, and Systems of Land Tenure." Journal of Political Economy 71 (February 1963): 1-29.

Nystuen, John, and Dacey, Michael. "A Graph Theory Interpretation of Nodal Regions." Papers, The Regional Science Association 7 (1961):29-42.

Otto, Fred. "Correlation of Harvests with Importation of Cereals in China." Chinese Economic Journal 15 (October 1934):388-414.

Perkins, Dwight H. "Government as an Obstacle to Industrialization: The Case of Nineteenth Century China." Journal of Economic History 27 (December 1967):478-92.

_____. "Growth and Changing Structure of China's Twentieth-Century Economy." In China's Modern Economy in Historical Perspectives, pp. 115-66. Edited by Dwight H. Perkins. Stanford: Stanford University Press, 1975.

"Production and Export of Groundnuts in China." Chinese Economic Journal 19 (September 1936):257-68; 19 (October 1936):374-95.

"Raw Cotton Trade in Shanghai." Chinese Economic Journal 3 (July 1928):681-92.

Rosenbaum, Arthur. "Railway Enterprise and Economic Development, the Case of the Imperial Railways of North China." Modern China 2 (April 1976):227-72.

Royal Asiatic Society. "Inland Communications in China." Chinese Economic and Social Life 2 (1890):1-213.

Shih, Min-hsiung. *The Development of the Silk Industry in China.* Translated by E-tu Zen Sun. Ann Arbor: University of Michigan, Center for Chinese Studies, 1976.

Skinner, G. William. "Marketing and Social Structure in Rural China, Part I." *Journal of Asian Studies* 24 (November 1964):3-44.

_____. "Marketing and Social Structure in Rural China, Part II." *Journal of Asian Studies* 24 (February 1965):195-228.

_____. "Regional Urbanization in Nineteenth-Century China." In *The City in Late Imperial China*, pp. 211-49. Edited by G. William Skinner. Stanford: Stanford University Press, 1977.

Spencer, J. E. "Trade and Transhipment in the Yangtze Valley." *Geographical Review* 28 (January 1938):112-23.

Spring, D. "Railways and Economic Development in Turkestan before 1917." In *Russian Transport: A Historical and Geographical Analysis*, pp. 46-74. Edited by L. Symons and C. White. London: Bell, 1975.

Sun, E-tu Zen. "The Pattern of Railway Development in China." *Far Eastern Quarterly* 14 (February 1955):179-99.

Swen, W. Y. "Types of Farming, Costs of Production, and Annual Labor Distribution in Wei-hsien, Shantung." *Chinese Economic Journal* 3 (July 1928):642-79.

Taaffe, Edward J.; Morrill, Richard L.; and Gould, Peter R. "Transport Expansion in Underdeveloped Countries: A Comparative Analysis." *Geographical Review* 53 (October 1963):503-29.

"Tea Production and Trade in Chekiang." *Chinese Economic Journal* 14 (May 1934):521-36.

Thomason, Conrad. "Ambiguities of Modernization: The Case of China." *The China Geographer* No. 5 (Fall 1976):13-27.

"Tobacco Production and Marketing in China." *Chinese Economic Journal* 12 (May 1935):407-20 (Part I); 531-43 (Part II).

Tsao, Line-an. "The Marketing of Soya Bean and Bean Oil in China." *Chinese Economic Journal* 7 (September 1930):941-71.

Tsha, Kyung-we. "Railway Transportation Conference." *Chinese Economic Journal* 9 (July 1931):729-38.

Tung, C. S. "A Review of Chinese Railways." *Chinese Economic Journal* 8 (January 1931):39-50.

White, C. "The Impact of Russian Railway Construction on the Market for Grain in the 1860s and 1870s." In *Russian Transport: A Historical and Geographical Analysis*, pp. 1-45. Edited by L. Symons and C. White. London: Bell, 1975.

Wong, John. "Peasant Economic Behavior: The Case of Traditional Agricultural Co-operation in China." *The Developing Economies* 9 (June 1971):332-49.

Zarembka, Paul. "Marketable Surplus and Growth in the Dual Economy." *Journal of Economic Theory* 2 (June 1970):107-21.

Zellner, Arnold: Kmenta, J.; and Dreze, J. "Specification and Estimation of Cobb-Douglas Production Function Models." *Econometrica* 34 (October 1966):784-95.

THE UNIVERSITY OF CHICAGO
DEPARTMENT OF GEOGRAPHY
RESEARCH PAPERS (Lithographed, 6×9 inches)

Available from Department of Geography, The University of Chicago, 5828 S. University Avenue, Chicago, Illinois 60637, U.S.A. Price: $8.00 each; by series subscription, $6.00 each.

LIST OF TITLES IN PRINT

48. BOXER, BARUCH. *Israeli Shipping and Foreign Trade.* 1957. 162 p.
62. GINSBURG, NORTON, editor. *Essays on Geography and Economic Development.* 1960. 173 p.
71. GILBERT, EDMUND WILLIAM *The University Town in England and West Germany.* 1961. 79 p.
72. BOXER, BARUCH. *Ocean Shipping in the Evolution of Hong Kong.* 1961. 108 p.
91. HILL, A. DAVID. *The Changing Landscape of a Mexican Municipio, Villa Las Rosas, Chiapas.* 1964. 121 p.
101. RAY, D. MICHAEL. *Market Potential and Economic Shadow: A Quantitative Analysis of Industrial Location in Southern Ontario.* 1965. 164 p.
102. AHMAD, QAZI. *Indian Cities: Characteristics and Correlates.* 1965. 184 p.
103. BARNUM, H. GARDINER. *Market Centers and Hinterlands in Baden-Württemberg.* 1966. 172 p.
105. SEWELL, W. R. DERRICK, et al. *Human Dimensions of Weather Modification.* 1966. 423 p.
107. SOLZMAN, DAVID M. *Waterway Industrial Sites: A Chicago Case Study.* 1967. 138 p.
108. KASPERSON, ROGER E. *The Dodecanese: Diversity and Unity in Island Politics.* 1967. 184 p.
109. LOWENTHAL, DAVID, editor, *Environmental Perception and Behavior.* 1967. 88 p.
112. BOURNE, LARRY S. *Private Redevelopment of the Central City, Spatial Processes of Structural Change in the City of Toronto.* 1967. 199 p.
113. BRUSH, JOHN E., and GAUTHIER, HOWARD L., JR., *Service Centers and Consumer Trips: Studies on the Philadelphia Metropolitan Fringe.* 1968. 182 p.
114. CLARKSON, JAMES D., *The Cultural Ecology of a Chinese Village: Cameron Highlands, Malaysia.* 1968. 174 p.
115. BURTON, IAN, KATES, ROBERT W., and SNEAD, RODMAN E. *The Human Ecology of Coastal Flood Hazard in Megalopolis.* 1968. 196 p.
117. WONG, SHUE TUCK, *Perception of Choice and Factors Affecting Industrial Water Supply Decisions in Northeastern Illinois.* 1968. 93 p.
118. JOHNSON, DOUGLAS L. *The Nature of Nomadism: A Comparative Study of Pastoral Migrations in Southwestern Asia and Northern Africa.* 1969. 200 p.
119. DIENES, LESLIE. *Locational Factors and Locational Developments in the Soviet Chemical Industry.* 1969. 262 p.
120. MIHELIČ, DUŠAN. *The Political Element in the Port Geography of Trieste.* 1969. 104 p.
121. BAUMANN, DUANE D. *The Recreational Use of Domestic Water Supply Reservoirs: Perception and Choice.* 1969. 125 p.
122. LIND, AULIS O. *Coastal Landforms of Cat Island, Bahamas: A Study of Holocene Accretionary Topography and Sea-Level Change.* 1969. 156 p.
123. WHITNEY, JOSEPH B. R. *China: Area, Administration and Nation Building.* 1970. 198 p.
124. EARICKSON, ROBERT. *The Spatial Behavior of Hospital Patients: A Behavioral Approach to Spatial Interaction in Metropolitan Chicago.* 1970. 138 p.
125. DAY, JOHN CHADWICK. *Managing the Lower Rio Grande: An Experience in International River Development.* 1970. 274 p.
126. MacIVER, IAN. *Urban Water Supply Alternatives: Perception and Choice in the Grand Basin Ontario.* 1970. 178 p.
127. GOHEEN, PETER G. *Victorian Toronto, 1850 to 1900: Pattern and Process of Growth.* 1970. 278 p.
128. GOOD, CHARLES M. *Rural Markets and Trade in East Africa.* 1970. 252 p.
129. MEYER, DAVID R. *Spatial Variation of Black Urban Households.* 1970. 127 p.
130. GLADFELTER, BRUCE G. *Meseta and Campiña Landforms in Central Spain: A Geomorphology of the Alto Henares Basin.* 1971. 204 p.
131. NEILS, ELAINE M. *Reservation to City: Indian Migration and Federal Relocation.* 1971. 198 p.
132. MOLINE, NORMAN T. *Mobility and the Small Town, 1900–1930.* 1971. 169 p.
133. SCHWIND, PAUL J. *Migration and Regional Development in the United States.* 1971. 170 p.

134. PYLE, GERALD F. *Heart Disease, Cancer and Stroke in Chicago: A Geographical Analysis with Facilities, Plans for 1980.* 1971. 292 p.
135. JOHNSON, JAMES F. *Renovated Waste Water: An Alternative Source of Municipal Water Supply in the United States.* 1971. 155 p.
136. BUTZER, KARL W. *Recent History of an Ethiopian Delta: The Omo River and the Level of Lake Rudolf.* 1971. 184 p.
139. McMANIS, DOUGLAS R. *European Impressions of the New England Coast, 1497–1620.* 1972. 147 p.
140. COHEN, YEHOSHUA S. *Diffusion of an Innovation in an Urban System: The Spread of Planned Regional Shopping Centers in the United States, 1949–1968,* 1972. 136 p.
141. MITCHELL, NORA. *The Indian Hill-Station: Kodaikanal.* 1972. 199 p.
142. PLATT, RUTHERFORD H. *The Open Space Decision Process: Spatial Allocation of Costs and Benefits.* 1972. 189 p.
143. GOLANT, STEPHEN M. *The Residential Location and Spatial Behavior of the Elderly: A Canadian Example.* 1972. 226 p.
144. PANNELL, CLIFTON W. *T'ai-chung, T'ai-wan: Structure and Function.* 1973. 200 p.
145. LANKFORD, PHILIP M. *Regional Incomes in the United States, 1929–1967: Level, Distribution, Stability, and Growth.* 1972. 137 p.
146. FREEMAN, DONALD B. *International Trade, Migration, and Capital Flows: A Quantitative Analysis of Spatial Economic Interaction.* 1973. 201 p.
147. MYERS, SARAH K. *Language Shift Among Migrants to Lima, Peru.* 1973. 203 p.
148. JOHNSON, DOUGLAS L. *Jabal al-Akhdar, Cyrenaica: An Historical Geography of Settlement and Livelihood.* 1973. 240 p.
149. YEUNG, YUE-MAN. *National Development Policy and Urban Transformation in Singapore: A Study of Public Housing and the Marketing System.* 1973. 204 p.
150. HALL, FRED L. *Location Criteria for High Schools: Student Transportation and Racial Integration.* 1973. 156 p.
151. ROSENBERG, TERRY J. *Residence, Employment, and Mobility of Puerto Ricans in New York City.* 1974. 230 p.
152. MIKESELL, MARVIN W., editor. *Geographers Abroad: Essays on the Problems and Prospects of Research in Foreign Areas.* 1973. 296 p.
153. OSBORN, JAMES F. *Area, Development Policy, and the Middle City in Malaysia.* 1974. 291 p.
154. WACHT, WALTER F. *The Domestic Air Transportation Network of the United States.* 1974. 98 p.
155. BERRY, BRIAN J. L., et al. *Land Use, Urban Form and Environmental Quality.* 1974. 440 p.
156. MITCHELL, JAMES K. *Community Response to Coastal Erosion: Individual and Collective Adjustments to Hazard on the Atlantic Shore.* 1974. 209 p.
157. COOK, GILLIAN P. *Spatial Dynamics of Business Growth in the Witwatersrand.* 1975. 144 p.
159. PYLE, GERALD F. et al. *The Spatial Dynamics of Crime.* 1974. 221 p.
160. MEYER, JUDITH W. *Diffusion of an American Montessori Education.* 1975. 97 p.
161. SCHMID, JAMES A. *Urban Vegetation: A Review and Chicago Case Study.* 1975. 266 p.
162. LAMB, RICHARD F. *Metropolitan Impacts on Rural America.* 1975. 196 p.
163. FEDOR, THOMAS STANLEY. *Patterns of Urban Growth in the Russian Empire during the Nineteenth Century.* 1975. 245 p.
164. HARRIS, CHAUNCY D. *Guide to Geographical Bibliographies and Reference Works in Russian or on the Soviet Union.* 1975. 478 p.
165. JONES, DONALD W. *Migration and Urban Unemployment in Dualistic Economic Development.* 1975. 174 p.
166. BEDNARZ, ROBERT S. *The Effect of Air Pollution on Property Value in Chicago.* 1975. 111 p.
167. HANNEMANN, MANFRED. *The Diffusion of the Reformation in Southwestern Germany, 1518–1534.* 1975. 248 p.
168. SUBLETT, MICHAEL D. *Farmers on the Road. Interfarm Migration and the Farming of Noncontiguous Lands in Three Midwestern Townships, 1939–1969.* 1975. 228 pp.
169. STETZER, DONALD FOSTER. *Special Districts in Cook County: Toward a Geography of Local Government.* 1975. 189 pp.
170. EARLE, CARVILLE V. *The Evolution of a Tidewater Settlement System: All Hallow's Parish, Maryland, 1650–1783.* 1975. 249 pp.
171. SPODEK, HOWARD. *Urban-Rural Integration in Regional Development: A Case Study of Saurashtra, India—1800–1960.* 1976. 156 pp.

172. COHEN, YEHOSHUA S. and BERRY, BRIAN J. L. *Spatial Components of Manufacturing Change.* 1975. 272 pp.
173. HAYES, CHARLES R. *The Dispersed City: The Case of Piedmont, North Carolina.* 1976. 169 pp.
174. CARGO, DOUGLAS B. *Solid Wastes: Factors Influencing Generation Rates.* 1977. 112 pp.
175. GILLARD, QUENTIN. *Incomes and Accessibility. Metropolitan Labor Force Participation, Commuting, and Income Differentials in the United States, 1960–1970.* 1977. 140 pp.
176. MORGAN, DAVID J. *Patterns of Population Distribution: A Residential Preference Model and Its Dynamic.* 1978. 216 pp.
177 STOKES, HOUSTON H.; JONES, DONALD W. and NEUBURGER, HUGH M. *Unemployment and Adjustment in the Labor Market: A Comparison between the Regional and National Responses.* 1975. 135 pp.
179. HARRIS, CHAUNCY D. *Bibliography of Geography. Part I. Introduction to General Aids.* 1976. 288 pp.
180. CARR, CLAUDIA J. *Pastoralism in Crisis. The Dasanetch and their Ethiopian Lands.* 1977. 339 pp.
181. GOODWIN, GARY C. *Cherokees in Transition: A Study of Changing Culture and Environment Prior to 1775.* 1977. 221 pp.
182. KNIGHT, DAVID B. *A Capital for Canada: Conflict and Compromise in the Nineteenth Century.* 1977. 359 pp.
183. HAIGH, MARTIN J. *The Evolution of Slopes on Artificial Landforms: Blaenavon, Gwent.* 1978. 311 pp.
184. FINK, L. DEE. *Listening to the Learner. An Exploratory Study of Personal Meaning in College Geography Courses.* 1977. 200 pp.
185. HELGREN, DAVID M. *Rivers of Diamonds: An Alluvial History of the Lower Vaal Basin.* 1979. 399 pp.
186. BUTZER, KARL W., editor. *Dimensions of Human Geography: Essays on Some Familiar and Neglected Themes.* 1978. 201 pp.
187. MITSUHASHI, SETSUKO. *Japanese Commodity Flows.* 1978. 185 pp.
188. CARIS, SUSAN L. *Community Attitudes toward Pollution.* 1978. 226 pp.
189. REES, PHILIP M. *Residential Patterns in American Cities, 1960.* 1979. 424 pp.
190. KANNE, EDWARD A. *Fresh Food for Nicosia.* 1979. 116 pp.
191. WIXMAN, RONALD. *Language Aspects of Ethnic Patterns and Processes in the North Caucasus.* 1980. 224 pp.
192. KIRCHNER, JOHN A. *Sugar and Seasonal Labor Migration: The Case of Tucumán, Argentina.* 1980. 158 pp.
193. HARRIS, CHAUNCY D. and FELLMANN, JEROME D. *International List of Geographical Serials, Third Edition, 1980.* 1980. 457 p.
194. HARRIS, CHAUNCY D. *Annotated World List of Selected Current Geographical Serials, Fourth, Edition. 1980.* 1980. 165 p.
195. LEUNG, CHI-KEUNG. *China: Railway Patterns and National Goals.* 1980. 235 p.
196. LEUNG, CHI-KEUNG and NORTON S. GINSBURG, eds. *China: Urbanization and National Development.* 1980. 280 p.
197. DAICHES, SOL. *People in Distress: A Geographical Perspective on Psychological Well-being.* 1981. 199 p.
198. JOHNSON, JOSEPH T. *Location and Trade Theory: Industrial Location, Comparative Advantage, and the Geographic Pattern of Production in the United States.* 1981. 107 p.
201. LICATE, JACK A. *Creation of a Mexican Landscape: Territorial Organization and Settlement in the Eastern Puebla Basin, 1520-1605.* 1981. 143 p.
203. LIANG, ERNEST P. *China: Railways and Agricultural Development, 1875-1935.* 1982. 186 p.

DATE DUE